Ancient Hindu Science

Its Transmission and Impact on World Cultures

Ancient Hindu Science: Its Transmission and Impact on World Cultures
Alok Kumar

ISBN: 978-3-031-79401-8 paperback
ISBN: 978-3-031-79402-5 ebook
ISBN: 978-3-031-00148-2 hardcover

DOI 10.1007/978-3-031-79402-5

A Publication in the Springer series
SYNTHESIS LECTURES ON ENGINEERING

Lecture #34
Series ISSN
Print 1939-5221 Electronic 1939-523X

Ancient Hindu Science

Its Transmission and Impact on World Cultures

Alok Kumar
State University of New York at Oswego

SYNTHESIS LECTURES ON ENGINEERING #34

ABSTRACT

To understand modern science as a coherent story, it is essential to recognize the accomplishments of the ancient Hindus. They invented our base-ten number system and zero that are now used globally, carefully mapped the sky and assigned motion to the Earth in their astronomy, developed a sophisticated system of medicine with its mind-body approach known as Ayurveda, mastered metallurgical methods of extraction and purification of metals, including the so-called Damascus blade and the Iron Pillar of New Delhi, and developed the science of self-improvement that is popularly known as yoga. Their scientific contributions made impact on noted scholars globally: Aristotle, Megasthenes, and Apollonius of Tyana among the Greeks; Al-Birūnī, Al-Khwārizmī, Ibn Labbān, and Al-Uqlīdisī, Al-Jāḥiz among the Islamic scholars; Fa-Hien, Hiuen Tsang, and I-tsing among the Chinese; and Leonardo Fibbonacci, Pope Sylvester II, Roger Bacon, Voltaire and Copernicus from Europe. In the modern era, thinkers and scientists as diverse as Ralph Waldo Emerson, Johann Wolfgang von Goethe, Johann Gottfried Herder, Carl Jung, Max Müller, Robert Oppenheimer, Erwin Schrödinger, Arthur Schopenhauer, and Henry David Thoreau have acknowledged their debt to ancient Hindu achievements in science, technology, and philosophy.

The American Association for the Advancement of Science (AAAS), one of the largest scientific organizations in the world, in 2000, published a timeline of 100 most important scientific finding in history to celebrate the new millennium. There were only two mentions from the non-Western world: (1) invention of zero and (2) the Hindu and Mayan skywatchers astronomical observations for agricultural and religious purposes. Both findings involved the works of the ancient Hindus.

The Ancient Hindu Science is well documented with remarkable objectivity, proper citations, and a substantial bibliography. It highlights the achievements of this remarkable civilization through painstaking research of historical and scientific sources. The style of writing is lucid and elegant, making the book easy to read. This book is the perfect text for all students and others interested in the developments of science throughout history and among the ancient Hindus, in particular.

KEYWORDS

Hindu science, History of science, Vedic science, Hindu religion, Ancient Indian science, Indian science and technology

This book is dedicated to my parents,
Late Ganga Saran Sarswat, father
and
Late Shanti Devi, mother.
They taught me virtues of life.

Contents

Preface

I was raised in Haridwar, a famous Indian city that is known for religion, philosophy, mysticism, and the Ganges river. I heard about the greatness of India often as a child from my father who was a learned man. I did not notice about this greatness in the scientific literature that was a part of my academic curricula. It created a great emotional dilemma to me. Why India, with so much philosophy, intellect, and prosperity, could not make a substantial contribution to science? I found the answer only after I came to America and had access to good library facilities in California. The history of science as we know from the textbooks is simply incomplete. By writing this book and other books, I am trying to fill these gaps.

It is not possible to provide details of the achievements of the ancient Hindus in one introductory book. Their contributions are enormous and this book presents only the 'tip of the iceberg,' as the phrase goes. I have chosen only those topics that are interesting to me and I have some knowledge.

The accomplishments of the ancient Hindus span many fields. In mathematics, they invented our base-ten number system and zero that are now used globally, carefully mapped the sky and assigned motion to the Earth in their astronomy, developed a sophisticated system of medicine with its mind-body approach known as Ayurveda, mastered metallurgical methods of extraction and purification of metals, including the so-called Damascus blade and the Iron Pillar of New Delhi, and developed the science of self-improvement that is popularly known as yoga. Their scientific contributions made impact on noted scholars from all over the world, Aristotle, Megasthenes, and Apollonius of Tyana among the Greeks; Al-Birūnī, Al-Khwārizmī, Ibn Labbān, and Al-Uqlīdisī, Al-Jāḥiz among the Islamic scholars; Fa-Hien, Hiuen Tsang, and I-tsing among the Chinese; and Leonardo Fibbonacci, Pope Sylvester II, Roger Bacon, Voltaire and Copernicus from Europe. Their testimony about Hindu science provide a clear sense of the immense contributions of the ancient Hindus. In the modern era, thinkers and scientists as diverse as Ralph Waldo Emerson, Johann Wolfgang von Goethe, Johann Gottfried Herder, Carl Jung, Max Müller, Robert Oppenheimer, Erwin Schrödinger, Arthur Schopenhauer, and Henry David Thoreau have acknowledged their debt to ancient Hindu achievements in science, technology, and philosophy.

In this book, I have used scientific norms of analysis and have sorted out the hard facts from fantasy. In other words, the analysis here is rational and objective. For important statements, I have provided citations to the peer-reviewed literature. This can help the readers to investigate further, if needed.

No culture or civilization has prospered to great heights without knowing and preserving their historic and existing knowledge base. Preserving knowledge is a process in which all generations must participate otherwise the knowledge become prone to be lost forever. This is my mindset in writing this book.

After I was done with the manuscript of one of my previous books, *Sciences of the Ancient Hindus: Unlocking Nature in the Pursuit of Salvation*, I submitted it to an internationally renowned publisher. After about two years of review process, the publisher agreed to publish the book provided I drop the term Hindus from the title and replace with Indian. One reviewer warned me of "the deeply contested nature of the adjective Hindu and its association with a particular kind of nationalist politics" in India. This was prior to Narendra Modi's government in India. I have no involvement in Indian politics now nor I ever had one at any stage. I have lived most of my adult life in America. I rejected the suggestion since I wanted to be truthful.

It has been a challenging and rewarding experience for me to write this book. I hope the readers enjoy reading this book as much as I have enjoyed writing it. Only the readers can judge the validity of this endeavor.

Alok Kumar
March 2019

Acknowledgments

After I completed my book, *Sciences of the Ancient Hindus*, I told my wife that I would not be writing another book on this topic. I said so since writing a book is a long arduous journey. It was difficult for me to conduct research for the book in the absence of a network of collaborators and proper academic support. I had to work on that book during my off hours from the job and the family-life suffered in the process.

Much was changed after I published the above-mentioned book. My family and I realized that this book was not just another academic publication. The book struck a chord with the readers and we were inspired to observe it. It changed our mindset. As a result, when I was approached by the editors of Morgan and Claypool Publishers, I readily accepted their offer.

Any arduous task becomes much simpler with a network of capable people to assist. I would like to thank the following people for their assistance:

- My wife, Kiran Singh-Kumar, daughter, Aarti Kumar, and mother-in-law, Chaya Singh, provided me constant encouragement and assistance. They are the silent heroes in this project.

- My brother, Nand Kishore Sharma, sister-in-law, Bina Sharma, and sister, Pushpa Sharma, who take so much pride in knowing my achievements.

- Dr Ved Chaudhary, President of Educators' Society for Heritage of India (ESHA); Dr Deen Khandelwal, Founder-President, Hindu University of America; Dr Ambalavanar Somaskanda, a medical doctor from Rochester; and Dr. John Kares Smith, my colleague from SUNY Oswego, for reading the first draft of the book. They made corrections on the draft, provided valuable suggestions to improve the book, and, above all, provided encouragement.

- Chris Hebblethwaite, librarian, who tirelessly searched databases to collect relevant information for me. His office was my first stop when I could not find a specific information.

- Dr. John Zielinski, my colleague in the physics department, who often goes for long walks with me on campus. He was always a willing participant in any discussion related to this book.

- Editors, Jeanine Burke and Joel Claypool, for providing me with excellent tips for effective writing.

I have tried hard to avoid printing and scholarly mistakes. However, if some remain, please bring them to my notice (alok.kumar@oswego.edu). If you like the book, the credit goes to the people mentioned above. I am responsible for the errors.

Alok Kumar
March 2019

CHAPTER 1

Introduction

"The first nation (to have cultivated science) is Hind. This is a powerful nation having a large population, and a rich kingdom. Hind is known for the wisdom of its people. Over many centuries, all the kings of the past have recognized the ability of the Hindus in all branches of knowledge."[1] This was the conclusion that Ṣā'id al-Andalusī (1029–1070) made in his book, *Ṭabaqāt al-'Umam* (Book of the Categories of Nations), in 1068. Ṣā'id lived in Spain and compiled perhaps the first popular book on the global history of science. Ṣā'id analyzed the scholarly contributions of various nations, chose eight nations that were well versed in sciences, and ranked Hind at the top of the list. The people that were described in his book for their contributions to science are: the Hindus, the Persians, the Chaldeans, the Greeks, the Romans, the Egyptians, the Arabs and the Hebrews.

Ṣā'id was a Muslim, a historian of science, and a mathematician with interest in astronomy. Ṣā'id, his father, and his grandfather served as religious judges (*qāḍi*) in Spain. In his role as judge, Ṣā'id mastered the sciences of jurisprudence and law, implemented the Sharia law in resolving conflicts, served as a mediator, and also supervised and audited the public works. Obviously, such roles were entrusted to persons of repute and influence. One of his students, Azarquiel (Arzachel or Zarqālī), is known for the famed *Toledan Tables*. These astronomical tables were used to predict the movements of the Sun, Moon and planets relative to the fixed stars. In the view of Ṣā'id, "[t]he Hindus, as known to all nations for many centuries, are the metal (essence) of wisdom, the source of fairness and objectivity. They are peoples of sublime pensiveness, universal apologues, and useful and rare inventions." In giving examples of such rare inventions, Ṣā'id mentioned the disciplines of mathematics, astronomy, medicine, and the invention of chess.

Ṣā'id's book was quite popular during the medieval period. During the colonial period, however, the book lost its repute, its contents did not fit well with the colonial agenda, and it was conveniently forgotten. It was introduced to the English-speaking world in 1991.[2]

Ṣā'id was familiar with the contributions of the Egyptians, Greeks and Romans to science. Yet, while comparing the significance of their contributions to science, he chose Hind to be the top nation in science. This is in contrast to what we teach today in sciences. Did Ṣā'id commit an

[1] Salem and Kumar, 1991, p. 11. In the original manuscript, the same term, Hind, is used to define the geographical region and the people. In today's context, the medieval term Hind describes the present India, Pakistan, Bangladesh, Nepal, and Afghanistan, popularly also known as the Greater India.

[2] Salem and Kumar, 1991. I was reading the scientific literature produced during the medieval world while researching for my book, *Sciences of the Ancient Hindus*. I noticed that Ṣā'id's book was cited by several medieval scholars. I tried to acquire the book and did not succeed. This led to more efforts and finally the original Arabic version was acquired, authenticated and published with proper translation and annotations.

honest scholarly mistake by placing Hind, also popularly called Bharat, India, and Hindustan, at the top of his list? Was he the only scholar to rank Hind at the top among all other nations in science? What are the important contributions of the Hindus to science?

Further, in year 2000, many events were organized and some landmarks were set to celebrate the new millennium. The American Association for the Advancement of Science (AAAS) tried to compile a list of the top 100 scientific findings that made significant impacts on the world. It was a major undertaking where quite a few historians of science were involved. Only two discoveries were selected from the non-Western world: (1) invention of zero and (2) the astronomical observations of the Hindu and Mayan skywatchers for agricultural and religious purposes. Both findings involved the works of the ancient Hindus. Why did the Hindus invent zero as a mathematical entity? What was the connection of astronomical observations with agricultural and religious purposes? Did they make any interesting astronomical observations in the process? These are some of the questions that this book has tried to answer. After going through its pages, readers will be able to make their own judgments on these issues.

1.1 THE MULTICULTURAL SCIENCE

While covering the ancient and medieval periods, most science courses focus on inventions and discoveries from Greece and Europe. Students learn that scientific rational thinking originated with the Greeks around the seventh century BCE, and flourished there for about 800 years. Greek philosopher-scientists such as Thales (624–546 BCE), Pythagoras (562–500 BCE), Democritus (460–370 BCE), Hippocrates (460–370 BCE), Plato (427–347 BCE), Aristotle (386–322 BCE), and Archimedes (287–212 BCE) are responsible for most basic ideas in science. The period after the beginning of the Common era is defined as *the Dark Ages* (475–1000 CE) or *the Middle Ages* (475 CE to the Renaissance). The term, *the Dark Ages*, signifies the lack of intellectual and scientific activities in Europe. After the fourteenth century, the Europeans reacquainted themselves with the scientific tradition of the Greeks that led to the European Renaissance. In relation to the Renaissance period, we learn about Galileo, Faraday, Newton, Kepler, and Boyle who lived in Europe. There are not many examples of scientists, discoveries, or inventions that have any connection to Asia, Africa, and Latin America.

Science evolves out of human necessities and curiosities. With the growth in science, our lives are constantly changing/improving in myriad ways. The Earth that was considered to be boundless by our ancestors can now be traveled around in a day or two. We have landed on the Moon and are planning our visit to Mars. We can easily make a telephone call to our loved ones halfway around the Earth for a nominal expense. The increased food demand in the past century is met by the green revolution. The life expectancy is increasing all over the world. Most civilizations in the past have found material benefits and intellectual satisfaction in attempting to understand the world's physical and biological phenomena. Science was bound to prosper in most cultures. The question is why it did not happen in Asia, Africa, the Middle East, and Latin America. Or, may be our science textbooks are simply providing incomplete information.

Indeed, this popular version of our history of science is full of gaps that are finally catching the attention of scholars. A major gap was first demonstrated by Joseph Needham (1900–1995) with his multi-volume book, *Science and Civilisation in China.* Needham was a British biochemist and historian who raised the famous question, popularly known as the Needham question: "Why did modern science, the mathematization of hypotheses about Nature, with all its implications for advanced technology, take its meteoric rise only in the West at the time of Galileo [but] had not developed in Chinese civilisation or Indian civilisation?" Needham answered this question in his book with a specific focus on China and proved that the history of science that we teach in science courses is simply incomplete. No such major effort to compile Indian science has taken place so far for a variety or reasons. The present book is a small effort to fill that gap.

Roger Bacon (1214–1294), a noted Franciscan natural philosopher from England, wrote a book, *The Opus Majus*, under the instruction of Pope Clement IV (1190–1268). The main purpose of the book was to improve the training of missionaries to Christianize distant ethnic lands.[3] This book clearly establishes that India was a leader in science. Bacon knew the works of Ibn al-Haytham (Alhazen)[4] (965–1040 CE), Al-Battānī (858–929 CE), and Ibn Sīnā (980–1037 CE) from the Middle East. He also knew about the Hindu science from his days in the University of Paris.[5]

Geoffrey Chaucer (c. 1343–in the Prologue of his *Canterbury Tales*[6], writes about a physician who was well versed with the works of Serapion the Elder of Syria, al-Rāzī and Ibn Sīnā of Persia, along with the works of Hippocrates, Rufus of Ephesus, Dioscorides and Galen:

> "With us ther was a Doctour of Phisyk
> In al this world ne was ther noon him lyk
> To speke of phisik and surgerye,

[3]Roger Bacon was not the only one who worked tirelessly to produce a book to assist the training of missionaries to Christinize India. Max Müller, Professor of Comparative Philology, Robert Boyle, Director of the East India Company, and Monier Monier-Williams, Boden Professor of Sanskrit in Oxford University, are some other noted scholars who produced literature or used their resources to assist missionaries to Christinize India. Monier-Williams even candidly wrote that the purpose of translation was to aid in "the conversion of the natives of India to the Christian religion." (Goldberg, 2010, p. 28.) Another person who made a significant impact to achieve this goal was Lord T. B. Macaulay (1800–1859), member of the Supreme Council of India. In this capacity, in his Minute on Indian Education, he suggested the British empire to introduce western-based reforms in Indian schools. This document became quite successful. Macaulay believed that (1) "a single shelf of a good European library was worth the whole native literature of India and Arabia" and (2) "all the books written in the Sanscrit [Sanskrit] language are less valuable than what may be found in the most paltry abridgments used at preparatory schools in England." With this mindset, the education policy of India was framed during the colonial period. Lord Macaulay's reforms largely remained in place in India even after the independence.

[4]The year 2015 was declared as the "Year of Light" by the United Nations to emphasize the importance of light science and to celebrate 1,000 years of Ibn al-Haytham's book, *Kitāb al-Manazir*, a book on optics. Several centuries later, many noted scientists, such as Roger Bacon, Robert Grosseteste, Leonardo Da Vinci, Galileo Galilei, Christiaan Huygens, René Descartes, Johannes Kepler, and Isaac Newton, had studied optics from a Latin translation or the original Arabic copy of his book. Some of them wrote their own books on optics later.

[5]Smith, a chapter, The Place of Roger Bacon in the History of Mathematics, in the book by Little, 1914, p. 156.

[6]*The Canterbury Tales*, Prologue, 411–413, 429–432. It is interesting to note the evolution of the English language in the past millennium.

Wel knew he the olde Esculapius,
And Deiscorides, and eek Rufus,
Old Ypocras, Haly, and Galien,
Serapion, Razis, and Avicen"

England, France, Spain, Portugal, and the Netherlands controlled most of the world during the eighteenth century. Popular literature was produced and disseminated by European nations to support their dream of domination. In this literature, Egypt, India, Persia, and China entered the scientific age only through their interactions with the Europeans. Thus, the histories of Asians, Africans, and other indigenous peoples often appear only after their encounter with the Europeans.[7]

Though the British and French governments were brainwashing these ethnic civilizations with their propaganda, they were recruiting their best scholars and scientists to learn from these ethnic cultures. For example, when Napoleon Bonaparte (1769–1821) invaded Egypt, he took about 150 biologists, mineralogists, linguists, mathematicians, and chemists with him to learn Egyptian science. This group included mathematician Jean-Baptiste Joseph Fourier, mineralogist Déodat Guy Grater de Dolomieu, and botanical artist Pierre Joseph Redouté.[8] Why did Napoleon take the best scientists of France to Egypt in a war? Napoleon had brought 400 ships, 40,000 soldiers, and 167 scientists, engineers, and artists to Alexandria. In less than a month, he lost to the British soldiers led by Admiral Horatio Nelson. Yet his mission was a magnificent triumph. The savants (as French scientists and philosophers were called) had uncovered an invaluable treasure of relics in Egypt, including the Rosetta Stone, and established the Institut d'Égypte in Paris, the first institution in the world devoted to Egypt's ancient culture.

Thus, in Western literature, the history of Asians, Africans, or the indigenous peoples of the Americas often appears to begin only after their encounter with European people. Science is thus Eurocentric and incomplete in the process. This omission of the non-Western literature in the history of science is "deeply unjust to other civilizations. And unjust here means both untrue and unfriendly, two cardinal sins which mankind cannot commit with impunity," writes Joseph Needham.[9]

The failure to mention multicultural contributions in present-day science education has been noted by many scholars who have provided examples after examples of historical gaps in their books.[10] For example, the Greek philosopher Leucippus (born ca. 490 BCE) or his disciple Democritus (born 470 BCE) are generally credited for being the originators of the atomic theory. The contemporaneous Indian philosopher Kaṇāda, who lived sometime between

[7]This topic caught the attention of scholars during the later part of the twentieth century. However, considerably more research is needed to better understand the contributions of other civilizations. For more information, consult Baber, 1996; Goonatilake, 1984, 1992; and Said, 1978 and 1993.

[8]For more information, read Brier, 1999.

[9]Needham in Nakayama and Sivin, 1973, p. 1.

[10]Bernal, 1971, Harding, 1991, 1994; Needham, 1954–99; Rashed, 1996; and Teresi, 2002. This knowledge is yet to be incorporated appropriately in introductory science textbooks.

the sixth to tenth centuries BCE, is simply ignored in most science texts. Henry Margenau (1901–1997), a noted philosopher-physicist who served as Eugene Higgins Professor of Physics and Natural Philosophy at Yale University, pointed to this gap in his book, *Physics and Philosophy*, and wrote, "But the most remarkable feature . . . which I have never seen in American textbooks on the history of science is the atomic theory of philosopher Kanada [Kaṇāda]."[11] This is not an American issue, as listed by Henry Margenau; it is an academic issue that is global in scope. Kaṇāda's work is still not covered in science texts even in India, the region where he was born and lived. As mentioned earlier, the colonial education pattern established by Lord Macaulay still continues in India, a sad reflection of the still pervasive colonial mindset of Indian academia.

Although progress is being made, it is at a very slow pace. We have a long way to go to establish science as a truly global enterprise. A significant number of articles and books have been published in the last 25 to 30 years to add contributions from Islamic countries, including an *Encyclopedia of the History of Arabic Science*.[12] The scholarship on the Indian, Egyptian, and Mayan civilizations is still highly incomplete. Our new knowledge on Islamic science and Chinese science happened due to the large influx of money and human resources from China and the Middle East. In contrast, the governments from Latin American countries, India, and Egypt have not allocated much resources for this cause. We need another Joseph Needham(s) to raise resources and preserve the knowledge of these civilizations. The time is now.

India, Egypt, Baghdad, and Persia were centers of learning, along with Athens and Rome, during the ancient and/or early medieval periods. For example, the place-value system to write numbers that was invented by the Hindus is central to the growth of mathematics. Many of the medicinal treatments, surgical procedures, and anatomical knowledge came to the West from the *Caraka-Saṁhitā* and the *Suśruta-Saṁhitā* of India, the *Edwin Smith Surgical Papyrus*, the *Ebers Papyrus*, and the *Kahun Papyrus* of Egypt, and the *Book of Healing* and the *Canon of Medicine* of Ibn Sīnā of Persia. Similarly, the modern astronomy owes a lot to the works of Āryabhaṭa I (ca. 500) from India and al-Khwārizmī from Baghdad.

The Islamic influence on science is evident from the Arabic terms that are commonly used in science: alcohol, Aldebaran, algebra, almanac, alkali, algorithm, Altair, azimuth, Betelgeuse, calendar, Deneb, magazine, monsoon, nadir, ream, and zenith are all derived from the Arabic language. These words have become a part of the Western heritage and are listed in most English dictionaries. A similar list of Sanskrit words in English is provided in Chapter 9.

The three crucial inventions that influenced the modern world came from China: paper, gunpowder, and magnetic compass. Paper remained the most important tool for documentation for over a millennium after it was recently replaced by digital electronic technologies. Magnetic compass allowed traveler to navigate through an ocean while the gunpowder became a tool of conquest and subjugation after its use was discovered for making guns and cannons.[13]

[11]Margenau, 1978, p. XXX.

[12]Hogendijk and Sabra, 2003; Kennedy, 1970; King, 1983 and 1993; Kunitzsch, 1989 and 1983; Rashed, 1996; Saliba, 1994; Samsó, 1994; and Selin, 1997 and 2000.

[13]The list of such contributions is long, and is covered by Kumar, 2014; Montgomery and Kumar, 2015.

1.2 THE ANCIENT HINDU SCIENCE

As mentioned in the previous section, the modern place-value notation system (base 10) that is used to represent numbers, is Hindu in origin. In this system we write, for example, eleven as one and one, side by side (11). The one on the left is in the second place, as we count from the right to left. The magnitude of this one (1) on the left is equal to ten. Any number in this place has to be multiplied by ten. Practically all cultures invented their own number system: the Greeks, Romans, Egyptians, Babylonians, Mayans, and Chinese. The Greeks, the Romans, and the Egyptians did not use a place-value notational system although they did use base-10 system. In their systems, eleven was written as ten and one (for example, XI in the Roman system). The Hindu place-value notation system made it possible to write very large numbers and simplified mathematical calculations; therefore, it prevailed over the other systems. Just imagine reading the values of various stocks in a newspaper. It is easy to figure out that a number with three digits will be greater than a number with two digits. Similarly, a number with four digits will be greater than a number with three digits. It provides a quick comparison which is not possible with other systems. It also allows much faster arithmetical calculations. Nicolaus Copernicus (1473–1543) in his book, *On the Revolutions*, used Hindu numerals in mathematical computations to provide a heliocentric model of the solar system. He noted the usefulness of this system over the Roman or Greek numeral system for quick computations. Copernicus was not the first one to use Hindu numerals. Several hundred years before him, Leonardo Fibonacci (1170–1250) wrote a popular book, *Liber Abaci*, and introduced the Hindu methods of numeration and computation to the Western world. Ṣā'id al-Andalusī wrote about the presence of this numeral system in Spain during the eleventh-century. (Read Chapter 3).

Trigonometry deals with relationships involving lengths and angles of right-angled triangles. Such mathematical relationships are highly applicable in the disciplines of architecture, mathematics, physics, and astronomy. With the work of Āryabhaṭa I (ca. 500), trigonometry began to assume its modern form. He used the half chord of an arc and the radius of a circle to define the sine of an angle.[14] The origin of the subject of trigonometry and word "sin," used to define trigonometric function "sin" (pronounced as *sine*), can be traced to the Sanskrit language. (Read Chapter 3).

About a millennium before Copernicus, Hindu astronomer Āryabhaṭa I assigned motion to the earth. He considered motion of the planets and considered stars to be stationary. Āryabhaṭa I used the analogy of a boatman in a river, observing objects on the shore moving backward, to explain the apparent motion of the Sun and other stars. He explained the concept of relative motion many centuries before its more formal discussion by the noted Parisian scholar Nicholas Oresme in the fourteenth century.[15] Interestingly, Copernicus used more or less the same analogy of a boatman to explain the apparent motion of the Sun in his book, *On the Revolutions* (Book 1, Chapter 8). (Read Chapter 4).

[14]Clark, 1930; Kumar, 1994.

[15]Kumar and Brown, 1999.

The ancient Hindus defined the age of the universe to be of the order of billions of years. This large number assigned to the age of the universe intrigued Carl Sagan, a noted astrophysicist who is famed for his Cosmos TV series. He wrote: "The Hindu dharma is the only one of the world's great faiths dedicated to the idea that the Cosmos itself undergoes an immense, indeed an infinite, number of deaths and rebirths. It is the only dharma in which time scales correspond to those of modern scientific cosmology. Its cycles run from our ordinary day and night to a day and night of Brahma, 8.64 billion years long, longer than the age of the Earth or the Sun and about half the time since the Big Bang." (Read Chapter 4.)

The ancient Hindus defined standards for physical measurements of space (length), mass, and time. The smallest unit of time used in India was of the order of 10^{-4} second, as noted by al-Bīrūnī, the Islamic scholar who visited India during the eleventh century. These standards were prevalent in India at least 500–1000 years before al-Bīrūnī. Similarly, multiple standards of length were also defined and, in *Mārkaṇdaya-Purāṇa*, the size of an atom was defined of the order of 10^{-9} meter. Kauṭilya (4th century BCE) defined various lever arms and scale-pans for balances for different range of weights. Superintendents were assigned to stamp labels for different weight-standards for public use to prevent cheating. Traders who did not use these stamped standard weights in business transactions were fined. Kaṇāda, in his *Vaiśeṣika-sūtra*, defined the concept of atom while discussing the distinctive properties of different matters and considering infinite divisions of matters. (Read Chapter 5).

Caraka, Suśruta, and Kauṭilya documented chemical transformations where oxidation, reduction, calcination, distillation, and sublimation were explained. Caraka, in his *Caraka-Saṁhitā*, lists gold, copper, lead, tin, iron, zinc, and mercury in making drugs. Mining was a highly organized activity among the ancient Hindus. Kauṭilya defines the role of mining for a sound economy. The Iron Pillar near Qutub-Minar in New Delhi is a testimony to metal forging of the ancient Hindus. The pillar, although about 1600 years old and weathering the heat, humidity, and rain in the open air, is still rust-free. We only hope that the car manufacturers of today can learn this ancient technology from India to make better rust-free cars. Hardened steel was also produced to allow a warrior to enter in a battlefield without worrying about breaking or bending his sword. This steel, though invented in India, is popularly known as Damascus steel. The Europeans first learned of this process from Damascus where it was called "steel of India." King Poros of Sindh, a province of India (now in Pakistan), after receiving a gift of life from Alexander the Great, gave 6,000 pounds of steel as precious gift to Alexander. (Read Chapter 6).

Plants have life; they try to protect themselves from the predators or attract bees for pollination purposes. At a time when human activities are destroying the ecology, the ecological perspectives of the ancient Hindus are relevant where rivers, mountains, plants, and animals are deemed sacred. The principle of *ahiṁsā* as a moral principle and its consequences on global warming and world hunger are discussed in the chapter on biology. (Read Chapter 7).

The so-called plastic surgery and cataract surgery find its roots in the surgical skills of the ancient Hindus. The role of a doctor and a patient, the design of a hospital, the role of food and

and the quality of air and water were considered by the ancient Hindus. They emphasized the body-mind approach to medicine and evolved ayurveda or "science of life," a system of medicine. (Read Chapter 8).

Ralph Emerson, Henry Thoreau, Leonardo Fibonacci, Schrödinger, Tolstoy, Tesla, Goethe, Schopenhaur, Robert Oppenheimer, Brian David Josephson (known for Josephson junction in physics) are some of the leading western thinkers who studied the scholarly work of ancient Hindus and formed their own worldview based on it. Similarly, on the east of India, Xuanzang (also known as Hiuen Tsang), Faxian (also known as Fa-Hien or Fa-Hsien), Yijing (also known as I-Ching, I-Tsing) were the leading scholars in China. These scholars are as much known for their wisdom as their arduous journeys to India to collect scholarly books. Similarly, on the West of India, Al-Jāḥiz, (ca. 776–868 CE), al-Khwārizmī (ca.800–847 CE), al-Uqlīdisī (ca. 920–980 CE) and Ibn Labbān (ca. 971–1029 CE), all noted Islamic philosophers, are known as much for their scholarly activities as for their efforts to introduce Hindu wisdom to the Middle East. These Islamic scholars were quite honest in writing their books, as they should be, and openly acknowledged their gratitude to the Hindus. (Read Chapters 3, 4, and 9).

1.3 ABOUT THE BOOK

This book explains the religious, social and cultural contexts that allowed some distinctive inventions and discoveries in Hindu science. For the Hindus, the disciplines of physics, chemistry, mathematics, astronomy, and medicine were sacred. A mastery of either of these disciplines allowed a person to achieve *mokṣa* (liberation from the cycle of birth and death), the highest goal of life for any Hindu.

This book primarily deals with the ancient period when the sacred books of the Hindus were composed(*Vedas*, *Upaniṣads*, and *Purāṇas*) and ends after the work of Āryabhaṭa I, the fifth-century. Therefore, the works of Brahmagupta, Bhāskara, and Mādhava, all prominent Hindu natural philosophers, are not included. Later contributions are covered in a few selective cases to know the impact of the ancient period. Since the works of the ancient Hindus took many centuries to become known to the Arabs and then to the Europeans, the later accounts from these cultures are discussed. For example, al-Bīrūnī's work during the eleventh century is discussed in relation to Āryabhaṭa I's work. The works of Emerson, Thoreau, Schrödinger, and Oppenheimer are discussed in connection to Vedanta or the *Bhagavad-Gītā*. Similarly, the work of Jagdish Chandra Bose related to plants is discussed in relation to *Mahābhārata*, an epic, where life in plants is explained in detail. Copernicus' comment about the usefulness of the place-value notation and his work in astronomy are discussed, although the works were done during the sixteenth-century.

The religious philosophy of the ancient Hindus may have played a crucial role in the invention of zero. The ancient Hindus tried to explain the nature of God that is devoid of all attributes (*Nirguṇa-swarūpa* or *amūrta*). This religious and philosophical approach of attribute negation or nothingness (*śūnyatā*) led them to the mathematical entity of infinite and zero.

In Chapter 2, readers will also learn the role of *śāstrārtha* (debate or discussion on the meaning of sacred texts) in resolving personal, social, and religious disputes. Conflicts were resolved without any rancor or violence. Even marriages were arranged using this practice (*svayaṁvara*). Also, the ancient Hindu literature was mostly written as poetry in lucid stories that are rich in similes and metaphors in order to facilitate its memorization and oral transmission. As a result, despite the destruction of libraries in the Indian peninsula after the Islamic invasion, this knowledge remained intact to a large extent. Pythagoreans in Greece also practiced orality to preserve their knowledge, like the ancient Hindus.

"Why do they call it by this name?" This etymological question is often asked by curious students. The epistemology and origins of various words and concepts that are commonly used in science texts, such as the so-called Arabic numerals, the zero, the trigonometric function "sine," algebra, and algorithm, are explained in this book. Information on these developments demonstrates the migration of knowledge from one culture to another and helps the readers to understand the multicultural nature of science.

The worldwide use of the Hindu numerals is perhaps a great triumph of the ancient Hindus. Although Hindu numerals were not accepted at first, rivalries had ensued in Arabia and Europe, decrees were issued against their use; their practical merit and their usefulness in mathematical calculations finally established their supremacy and they gradually became prevalent worldwide. Leonardo Fibbonacci (1170–1250) is well known for Fibbonacci's Sequence. However, little is known about his gratitude to the ancient Hindus, as he had clearly acknowledged in *Liber Abaci*.

Kaṇāda's book, *Vaiśeṣika-Sūtra*, defines the nature of time and space, conservation of matter, gravitation, and the concept of the atom. Time is a commonly used word in our daily conversation. However, its nature is enigmatic and subtle. Readers will learn the subtleties in the concept of time in Chapter 5. They will also learn the concept of the atom, as suggested by Kaṇāda, and its comparison with Democritus' atom.

This book is written primarily for readers who are trained in the Western knowledge system and are interested in learning about Hindus' contributions to science. Many references are deliberately chosen from the primary sacred literature of the Hindus as well as authentic secondary sources from the Western sources. Such a selection is made to counter a general skepticism demonstrated by some Western scholars concerning Hindu accounts of their history. These Western scholars generally complain that scholars in the East tend to stretch their imaginations to suit their views and do not provide logical steps and facts when deriving their conclusions. A hallmark of this book is in the documentation of the scholarly comments and acknowledgements made by Greeks, Persians, Egyptians, Arabs, Chinese, and Europeans in support of the scientific achievements of the ancient Hindus. Salient features of this book are in providing cross-cultural perspectives and comparisons, and portraying a coherent picture of the scientific contributions of the ancient Hindus.

This book is not encyclopedic or compendious. It is not possible to achieve that in such a small introductory work since the ancient Hindu literature is vast. This book presents only the "tip of the iceberg," as the saying goes. I have chosen only those topics where my knowledge and interests lie.

Several Sanskrit and Hindi words are now commonly used in English and have entered in English dictionaries. However, their English spellings in some cases are not in accordance with the system of transliteration. As these words are in common use in the English-speaking world, using any other spelling might create confusion for a broad range of readers. For this reason, we have kept the popular usage in some cases. For example, the spelling of the word "ayurveda," as mentioned in most dictionaries, is incorrect; the proper spelling is āyurveda. For Sanskrit terms that are not present in most English dictionaries, diacritic marks have been retained. To keep the book readable, simple, and enjoyable to non-scholars in the field, an amalgamated system of popular spelling as well as proper spelling is used. It is a common practice among scholars dealing with non-English literature.

I have used scientific norms of analysis and have sorted out the hard facts from fantasy. In other words, the analysis here is rational and objective. In philosophy and mysticism, there are several areas where the ancient Hindu literature stands abreast with the later concepts in science, such as causality and duality (or dualism). But these concepts are not covered in this book, because the borderline between facts and opinion is hazy.[16] Duality, as defined in the concept of the Creator and the creation as two independent entities, in Hindu philosophy, is quite different from the de-Broglie's Wave-Particle Duality of matter. The parallelism between ancient theories and modern science is fascinating to read; but in many cases, this is where the connection ends. The works of noted Noble Laureate physicists such as Brian Josephson and Erwin Schrödinger do pose an interesting dilemma in which their beliefs, based on the ancient Hindu literature, played a significant role in their discoveries in science.[17]

In the Lawrence Livermore Laboratory in California, the Shiva Target Chamber is a 20-laser-beam facility to study laser fusion. This facility was constructed in 1977. Edward Teller, the father of the hydrogen bomb and a designer of this facility, explained the design of the chamber in the following words: "Laser light is brought in simultaneously from ten pipes on the top and ten pipes on the bottom. Compression and nuclear reaction occurs in a tiny dot at the middle of the sphere. Apparatus practically filling a whole building feeds the twenty pipes, or the arms of the god Shiva [Śiva]. According to Hindu Creed, Shiva [Śiva] had three eyes: two for seeing, and one (usually kept closed) to emit annihilating radiation. The Hindus obviously knew about

[16] *The Tao of Physics* by Fritzof Capra, *The Wu Li Masters* by Gary Zukav, and *Mysticism and the New Physics* by Michel Talbot are examples of such works. These books are bestsellers for their insights. These are scholarly works that brought together the disciplines of religion and science.

[17] Read Capra, 1980; Josephson, 1987; Restivo, 1978 and 1982; Schrödinger, 1964; Talbot, 1981; and Zukav, 1979.

lasers."[18] Is it really true that the Hindus "knew about lasers"? Perhaps not. In my mind, what Teller has mentioned is a mere conceptual notion of Hindus in an immensely powerful laser like light that can destroy every thing in an instant, like the third eye of Lord Śiva, but it cannot be cited as a historical fact or an established theory.

The term Hindu was commonly used in science texts in the last century.[19] The leading journals of science, like *The American Mathematical Monthly*, *ISIS*,[20] *Islamic Culture*, *Science*, and *The Mathematical Gazette* and many books used the term Hindu science in the past. However, such usage is less these days since some authors are concerned about the reaction of the readers.[21] For example, Philip Goldberg, in his bestseller book, *American Veda*, avoided the term Hindu because he was concerned that "many potential readers would misconstrue the nature of the book."[22] In my previous book, *Sciences of the Ancient Hindus*, a reviewer warned me of "the deeply contested nature of the adjective Hindu and its association with a particular kind of national politics" in India. I have ignored such concerns. I do not have any involvement with the politics of India, nor do I want to push any political agenda through this book. The *Vedas*, the *Upanishads*, and *Purāṇas* are the sacred books of the Hindus. I want my book to be based in truth.

Once the readers realize the truthfulness of my assertions, I am confident that such myopic criticisms will disappear. "A class in arithmetic would be pleased to hear about the Hindoos [Hindus] and their invention of the 'Arabic notation'," suggested Cajori.[23] "They will marvel at the thousands of years which elapsed before people had even thought of introducing into the numeral notation that Columbus-egg—the zero."[24] It is this Columbus-egg, the zero, that is captivating the historians of science in AAAS as they include zero among the top 100 scientific finds that made significant impact in human history, as discussed earlier.

[18]Teller, 1979, p. 216.

[19]Datta, 1927; Hammett, 1938; Herschel, 1915; Karpinski, 1912; Mukhopādhyāya, 1994; Ray, 1919 and 1956; Renfro, 2007; Royle, 1837; Saidan, 1965; Seal, 1915; Zimmer, 1948.

[20]It is a mere coincidence that this term is also associated with a terrorist organization. The journal ISIS is a premier journal of the history of science.

[21]I am pleased that there are no such concerns with Islam and Buddhism where a good number of new book titles are published every year with explicit mention of religion.

[22]Goldberg, 2010, p. 2–3.

[23]Cajori, 1980, p. 3. Florian Cajori (1859–1930) was a professor and the first chair in history of mathematics at the University of California at Berkeley. Many of his books on the history of mathematics are still a landmark.

[24]Cajori, 1980, p. 3. Columbus-egg, a term that Cajori related to the invention of zero, refers to a brilliant discovery or idea that looks simple after fact.

CHAPTER 2

The Building Blocks of Science

As explained in the first chapter, human necessities play an important role in the evolution of science. People set up their goals based on their personal needs or the needs of the society and achieve them with the help of science. What were the geographical, social, and religious conditions that allowed the growth of science among the ancient Hindus? How did the ancient Hindus preserve their scientific knowledge and transmit it to the following generations? How people who contributed to Hindu science were treated in their society? These are important questions that need to be explored to understand the social, cultural, and religious contexts of the development of science among the ancient Hindus.

The sciences of the ancient Hindus were an essential and integral part of their religious practices. An important tenet of Hinduism is in the transmigration of soul which is defined in so many ways in the *Vedas* and *Upaniṣads*. This doctrine tells us that the soul is immortal and it transmigrates or reincarnates from one life form to another.[1] Our deeds in this life decide our fate in the next life. Therefore, the notion of rebirth is tightly coupled with the notion of *karma* (action) that provides a great incentive toward leading a moral life minimizing wrong deeds. Pythagoras, Socrates, Empedocles, Plato, Plotinus, Apollonius, and other Pythagorean philosophers also believed in the transmigration and immortality of the soul.

"The soul is neither born, nor dies. This one has not come from anywhere, has not become anyone. Unborn, constant, eternal, primeval, this one is not slain when the body is slain. If the slayer think to slay, if the slain think of himself slain, both these understand not. This one slays not, nor is slain," suggests *Kathā-Upaniṣad*.[2]

"Either as a worm, or as a moth, or as a fish, or as a bird, or as snake, or as a tiger, or as a person, or as some other in this or that condition, he is born again here according to his deeds, according to his knowledge," suggests *Kaushītaki-Upaniṣad*.[3]

The goal of life is to avoid this cycle of birth and death and achieve liberation (*mokṣa*). As a mother nourishing her children, gives *capāti* (*chapāti* or *rotī*, a flat bread) and *dāl* (lentils) to one, and *khicaḍī* (a rice and lentil preparation) and yogurt to another; similarly, according to their needs, the Hindu religion offers four major choices to human rational minds, and allows individuals to choose their own path to liberation:

1. *Karma-yoga* (the Path of Action, Selfless Service to Humanity)

[1] *Atharvaveda*, 12: 2: 52.
[2] *Kathā-Upaniṣad*, 2: 18, 19.
[3] *Kaushītaki-Upaniṣad*, 1: 2.

2. *Bhakti-yoga* (the Path of Devotion or Love to God)
3. *Jñāna-yoga* (the Path of Knowledge of Ultimate Truth)
4. *Rāja-yoga* (the Path of Yoga and Meditation)

All paths are equally effective. It is up to an individual to select the appropriate path for himself or herself. *Jñāna-yoga* (*Jñāna* means knowledge), the path relevant here, encourages an individual to understand the *ultimate truth* by raising such questions as, "Who am I?", "Why am I here?", or "What is the purpose of my life?" This mode of Socratic questioning allows self-introspection; the person discovers the ultimate truth from within. Like Socrates suggested in Greece, the *Bhagavad-Gītā* tells us that one can find knowledge from within and perfect it by yoga: "There is no better means of achieving knowledge; in time one will find that knowledge within oneself, when one is oneself perfected by yoga."[4] "We contemplate that adorable glory of the deity, that is in the earth, the sky, the heaven! May He stimulate our mental power."[5] This hymn is the popular *Gāyatri-mantra* that is chanted over and over by millions of Hindus as a ritual prayer every day.

The mind is the primal source of our knowledge. Our five senses would not function without the mind. It is the mind that unravels most of the obstacles that we face from day to day. One can even achieve liberation (*mokṣa*) through the exercise of mind to comprehend the ultimate truth. For this reason, knowledge has always remained so central to the Hindus. If knowledge is the key to salvation, then the issue of ownership of knowledge and defining spatio-temporal attributes associated with knowledge become trivial issues. Therefore, the ancient Hindus did not care much to define the period or author of their knowledge. This practice contrasts greatly with the Western tradition. As Charles Eliot elucidates: "They [Hindus] simply ask, is it true, what can I get from it? The European critic, who expects nothing of the sort from the work, racks his brain to know who wrote it and when, who touched it up and why?"[6] Hindu tradition has subordinated the pride of authorship, invention, or discovery to the self-satisfaction one gains from discovering the truth and sharing it with the world in the spirit of selfless service to humanity. While this avoids the cult of personality, it results in a lack of chronological records of discoveries and inventions. This explains why Hindus had developed such a vast literature yet there is no chronological records, so valued by the Western historians.

The ancient Hindus did not separate the disciplines of astronomy, mathematics, chemistry, physics, yoga, and medicine from moral codes, prayers, and the so-called divine literature. These scientific disciplines were labeled as sacred disciplines, necessary to know the ultimate knowledge. The *Chāndogya-Upaniṣad* cites an incidence in which the vagabond saint, Nārada, approaches another sage, Sanatkumāra, to learn about the ultimate knowledge—a knowledge

[4] *Bhagavad-Gītā*, 4: 38.

[5] *Ṛgveda*, 3: 62: 10.

[6] Eliot, 1954, vol. 1, p. LXVII. Charles Norton Edgecumbe Eliot (1862–1931) was a botanist, linguist, and diplomat. He was the British ambassador to Japan and was fluent in 16 languages and could converse in another 20 languages, including Sanskrit.

that could provide him freedom from *saṁsāra* (worldly manifestation) and lead to liberation (*mokṣa*). As all good teachers do, Sanatkumāra asked Nārada to apprise his existing knowledge base, so that an appropriate lesson can be designed. Nārada pointed out astronomy (*Nakṣatra-vidyā*) and mathematics (*rāsi-vidyā*), along with logic, history, grammar, fine arts, and the four *Vedas* as the knowledge that he had already mastered in his efforts to achieve *mokṣa*.[7]

In Hindu tradition, secular knowledge (*aparā-vidyā*) is considered to be helpful in achieving liberation, along with spiritual knowledge (*parā-vidyā*), as advised in *Muṇḍaka-Upaniṣad*.[8] When Śaunaka, a seeker, went to Aṅgirās, a teacher, and asked: "By knowing which a man comes to know the whole world." Aṅgirās' reply included a long list of disciplines, such as astronomy, the sacred *Vedas*, Sanskrit grammar, etymology and metrics, as suggested in *Muṇḍaka-Upaniṣad*.[9]

The natural philosophy of Hindus fulfills the spiritual needs of people as well as their need for rational thinking. It is for this reason that Ārybhaṭa I, in his book *Āryabhaṭīya*, covering astronomy, mathematics, and physics, suggested that one can achieve Brahman (*mokṣa*) by becoming well versed in the disciplines of astronomy, physics, and mathematics.[10] Similarly, Kaṇāda, in his book *Vaiśeṣika-sūtra*, defined the physical properties of matter and suggested that the knowledge is helpful in achieving *mokṣa*.[11] Also, the *Agni-Purāṇa* suggests that knowledge of the human anatomy can also lead to *mokṣa*. "Said the God of Fire: Now I shall describe the system of veins and arteries [*Nāḍī-cakra*] that are to be found in the human body. A knowledge of these [arteries and veins] leads to a knowledge of the divine Hari [God]."[12]

Scientific activities had important functions that were valued in the ancient Hindu society. For example, the role of astronomers was to fix the calendar, to set dates of religious festivals, and to predict eclipses and other astronomical events. These disciplines and duties became as important as composing and promoting moral codes. Learning human anatomy and functions helped in treating diseases in people and animals.

Knowing science had another important function: One way to know about the Creator (God) is to learn about God's creation. Science is an important tool to learn the physical properties of the created universe. Creation is the physical phenomenon that can tell us about the Creator. For this reason, Albert Einstein once wrote: "I maintain that the cosmic religious feeling is the strongest and noblest motive for scientific research. Thus, science became a tool to learn about the God. . . in this materializtic age of ours the serious scientific workers are the only profoundly religious people."[13]

[7] *Chāndogya-Upaniṣad*, 7: 1: 2–4.
[8] *Muṇḍaka-Upaniṣad*, 1: 1: 3–5.
[9] *Muṇḍaka-Upaniṣad*, 1: 1: 5
[10] *Āryabhaṭīya*, Daśgītika, 13. "Whoever knows *Daśgītika Sūtra* [ten verses] which describes the movements of the Earth and the planets in the sphere of the asterisms passes through the paths of the planets and asterisms [stars] and goes to the higher Brahman [God]."
[11] *Vaiśeṣika-Sūtra*, 1: 1: 4.
[12] *Agni-Purāṇa*, 214: 1–5.
[13] Einstein, 1930.

Progress in science has never been a hindrance to spiritual growth in the history of Hinduism. Knowing the truth was the main focus. Thus, science could grow freely and independently and no artificial boundaries within the moral codes limited the scientists in their investigations. This led al-Mas'udī (d. 957 CE), an Islamic historian during the tenth century, to write that science and technology were established without the aid of religious prophets in India. In his opinion, wise men could deduce the principles without the need of religion. It is not the prophets who dictated the domain of science, it was the logic, intuition, and experience of diligent observers who contributed to the domain of science. Al-Mas'udī considered Hind (India) as the land of "virtue and wisdom."[14] This is in contrast to prolong periods in some parts of the world where a chasm existed between science and religion and people had to make a choice between the two. It is well known that Bruno and Michael Servetes were burned to death and Galileo was imprisoned when their scientific beliefs were in conflict with the religious doctrines during the Inquisition period in Europe.

The lofty heights reached by the ancient Hindus in the realm of philosophy and religion are well recognized and extensive literature exists on these topics.[15] However, not much is known in the popular literature about their contributions to the natural sciences. The sciences of the ancient Hindus are embedded in their religious books, along with other disciplines. In ancient India, the various domains of knowledge, including science and religion, progressed hand-in-hand and grew under the shelter of one another. Religion flourished with the help of science and science flourished with the multi-faceted development of religion.

2.1 GEOGRAPHY

The current boundaries of India lie in the Northern Hemisphere between 8°4′N to 37°6′N latitudes and from 68°7′ E to 97°25′ E longitudes. Thus, the latitudinal as well as the longitudinal extent of India is about 29 degrees. Although India accounts for only 2.4 per cent of the world's total land area, it sustains 16 percent of the world population. The Tropic of Cancer (23°30′ N) divides the country into two equal halves: (1) the southern half lies in the tropical zone while the northern half belongs to subtropical zone. India has the Himalayan mountain range in the north, Vindhya mountain range in the middle, Indian Ocean in the south, Thar desert and Punjab plain in the west, forested mountains in the north-east, and tropical and watershed region of the Indo-Gangetic Plain in the east. The high Himalayan mountain range, with the tallest peak of mount Everest, blocks the frigid wind from the Tibetal Plateau and augurs temperate climate in the north. The south, in contrast, is always warm and humid. Thus, the geography of India provided all different kinds of climate.

With this geography comes vast mineral resources, and a great diversity of flora and fauna. It was easy for its inhabitants to meet their basic needs for food and shelter. The climate allowed a lifestyle where they could be close to the nature. The high population density in India is sus-

[14]Khalidi, 1975, p. 102–106.

[15]Dasgupta, 1922–1955; Durant, 1954; and Radhakrishnan, 1958.

tainable due to the favorable climatic conditions. Ralph Waldo Emerson (1803–1882 CE), an eminent American philosopher and poet, who lived in the northeast region of America where the temperature during the winter season is frigid, wrote: "The favor of the climate, making subsistence easy and encouraging an outdoor life, allows to the Eastern nations a highly intellectual organization, - leaving out of view, at present, the genius of Hindoos [Hindus] (more Orient in every sense), whom no people have surpassed in the grandeur of their ethical statement."[16] Emerson carefully studied the Hindu literature, including *Vedas*, *Upaniṣads*, *Bhagavad-Gītā*, and the works of Kālidāsa, a poet. Hindu philosophy was the source of Emerson of his transcendentalism and helped him in his quest to define a truly representative man.

Ancient India had acquired great fame as a society rich in spiritual as well secular dimensions. Therefore, it was visited by many travelers from Greece, Rome, China, and Arabia during the ancient and early medieval periods who provided accounts of the prosperity in the region. An early account is from Megasthenes (350–290 BCE), an ambassador of Seleucus I who ruled India for a short period. Megasthenes writes that the Indians, "having abundant means of subsistence, exceed in consequence the ordinary stature, and are distinguished by their proud bearing. They are also found to be well skilled in the arts, as might be expected of men who inhale a pure air and drink the very finest water."[17] Megasthenes is emphatic that the region never suffered from "famine" and "general scarcity in the supply of nourishing food."[18] Strabo (63 BCE - 21 CE), a Greek geographer and traveler, also visited India and found that the country was "abounding in herbs and roots,"[19] indicating prosperity.

The Chinese traveler Yijing (or I-tsing; 643–713 CE), who lived in India for about 22 years, mostly in Nalanda, a center of learning in modern Bihar, wrote: ". . . ghee [clarified butter], oil, milk, and cream are found everywhere. Such things as cakes and fruit are so abundant that it is difficult to enumerate them here."[20] Yijing visited India to acquire knowledge and carried some 400 Sanskrit texts back with him to China.[21]

Most Chinese, Greek, Persian, Arabian, and European documents written in different periods testify to the prosperity of the Indus-Sarsvatī Valley region. It is only after the later part of the British occupation, in the late nineteenth century that the region suffered from hunger and poverty due to colonial exploitation.

2.2 THE POWER OF QUESTIONING: *ŚĀSTRĀRTHA* (DEBATE) TO ACQUIRE KNOWLEDGE

Curiosity raises questions and questions lead to ideas and creativity. The power of questioning as a learning tool goes far back into history. The Socratic dialog is a standard tool for learning where

[16] Emerson, 1904, vol. 8, p. 239. For more information, read Acharya, 2001.
[17] McCrindle, 1926, p. 30.
[18] McCrindle, 1926, p. 31.
[19] Strabo, 15: 1: 22
[20] Takakusu, 1966, p. 44.
[21] Takakusu, 1966, p. xvii.

the answer of a question is a question. Socrates (470–399 BCE) didn't give lectures or write books. He propagated a dialectical technique in which he led his students by asking appropriate questions that demanded critical thinking to arrive at the correct answer. Socrates' questions revealed the ways in which his student's thinking was dogmatic and in error.

A good question is an excellent way to start a conversation. Good questions lure people to open up about themselves and divulge their thoughts and feelings. Questions are also instrumental in allowing you to introspect yourself and find answers. In a group, questions promote discussion. Questions have the ability to spawn more questions. This process is a hallmark of learning and has been an essential ingredient of Hindu thought.

In some religions, unquestioning faith in their scriptures is emphasized and used as the yardstick to judge the person's spirituality. In contrast, in Hinduism, asking questions is a common norm. The *Bhagavad-Gītā*, a sacred book of the Hindus, is a narrative dialog between prince Arjuna and Lord Kṛṣṇa. Arjuna was not fearful in asking tough questions; he was not fearful of being defined as a person with no faith. He bluntly asked, "Why should I fight with my own family members?" Kṛṣṇa, by fulfilling Arjuna's curiosity, could teach the concepts of *dharma*, duty, and *mokṣa*. The *Bhagavad-Gītā* is not the only book with such dialogs in the Hindu corpus; Maitreyī, in *Bṛhadāraṇyaka-Upaniṣad*, raises question about the futility of wealth and love for self. The sacred *Rāmāyaṇa* is the compilation of questions asked by sage Vālmīki and answers by Nārada Muni.

Henry David Thoreau, (1817–1862 CE), an American philosopher, naturalist, and author, praised the Hindus for their openness to new ideas. "The calmness and gentleness with which the Hindoo [Hindu] philosophers approach and discourse on forbidden themes is admirable."[22] Thoreau wrote this statement after his studies of the *Vedas* and *Upaniṣads*. Truthfulness, goodness, and beauty, as marked in *satyaṁ*, *śivaṁ*, and *sundaraṁ* [Only truth is beneficial and beautiful], have always been the guiding principles for the Hindus.

Interrogation, cross-examination, debate, symposium, and discussion were well defined tools practiced from the ancient period among the Hindus. In the Hindu tradition, scholastic debate (*vāda*) is practiced not only for the disciplines of philosophy; it is also for sciences and religion. Questioning is a powerful tool of investigation to discover unchartered territory of knowledge. Usually multiple possibilities are probed when people try to resolve a question. How do you select one possible solution over the others? How do you decide, when different people support different solutions? A debate is a way in which scholars can present their cases; such debates can easily filter novices from scholars. The practice of debate was so ingrained and valued among the Hindus that it was used as one of the eight ways in which a woman could select a groom for herself. Prospective grooms debated with established scholars or among a group of prospective grooms in public on various issues to win the bride. It was this practice that led

[22] Thoreau, 1906, vol. 2, p. 3.

Gautama (who later became Lord Buddha) to debate Arjuna, as mentioned in *Lalitvistara*, an ancient book (See Chapter 3).[23]

The *Bṛhadāraṇyaka-Upaniṣad* narrates an episode in which King Janaka decided to donate one thousand cows to the best Brahmin. Yājñavalkya, a revered saint, took all the cows and thus infuriated several other holy men who also needed the gift for their livelihood. To resolve the matter, a debate ensued and Yājñavalkya had to demonstrate his superior intellectual abilities by answering questions posed by other holy men.[24] Sages like Aśvala, Jāratkārava Ārtabhāga, Bhujyu Lāhyāyani, Uṣhasta Cākrāyaṇa, and Kahola Kauṣītakeya debated with him and asked questions on philosophical subjects to which Yājñavalkya provided convincing replies. They all lost the debate one-by-one. These sequence of debate was chosen by the rank of these scholars. In the end came Gārgi Vācaknavī, the daughter of sage Vācaknu who was in the lineage of sage Gārga, and took her name from the lineage and her father. But, she also lost to Yājñavalkya.[25]

A healthy *śāstrārtha* (debate or discussion) is an essential element and practice in the Hindu religion. It has led to the idea of "monism"—there is only one existence (God)—as well as the idea of "dualism"—there are two separate realities, Him (God) and me (my soul). The quest to know the ultimate truth led to the evolution of various systems of knowledge which outwardly seem to be divergent. The Hindu tradition allowed divergent opinions to coexist and established *śāstrārtha* as a means to resolve scholarly differences. Each person was allowed to test and discover the truth in his/her own way. Debate was considered a good way to effectively formulate the thought process, a good way to understand the subject matter, and to become established among scholars. Forty-four different forms of debates or discussions are described in the *Caraka-Saṁhitā*, demonstrating debate as a highly evolved system.[26] This book advises to decide the purpose of the debate in advance: For example, is it for curiosity or to subjugate the opponent? The latter purpose of subjugation was practiced in the special situations, e.g., *śāstrārtha* for marriage.

The Sanskrit term *ānuvīkṣikī*, defined as investigation through reasoning, has a long tradition among the ancient Hindus. It is a tool of investigation that is applicable in all aspects of learning: scientific, religious, and social. The earliest text on economics, *Arthaśāstra* of Kauṭilaya, lists[27] *ānuvīkṣikī* as one of the four cognitive (*vidyā*) disciplines, along with *trayī* (vedic learning), *daṇḍanīti* (jurisprudence), and *vārttā* (economics). *Ānuvīkṣikī* is considered as a "source of all knowledge" or a "means for all activities," and a "foundation for all social and religious duties." "When seen in the light of these sciences, the science of *ānuvīkṣikī* is most beneficial to the world, keeps the mind steady and firm in weal and woe alike, and bestows excellence of foresight, speech, and action. Light to all kinds of knowledge, easy means to accomplish all

[23] Bays, 1983, p. 224.

[24] *The Bṛhadāraṇyaka-Upaniṣad*, 3: 1.

[25] In contrast to many cultures, females could excel in philosophy and science among the ancient Hindus. The practice continues even today.

[26] *Caraka-Saṁhitā*, *Vimānasthāna*, 8: 27.

[27] King, Richard, 1999, p. 34; *Arthaśāstra*, 1: 2: 6–7.

kinds of acts and receptacle of all kinds of virtues, is the science of *ānuvīkṣikī* ever held to be," suggests Kauṭilaya.[28]

The medical treatise *Caraka-Saṁhitā* emphasizes the importance of debate and discussion in the learning process. "Discussion with a person of the same branch of science increases knowledge and brings happiness. It contributes toward the clarity of understanding, increases dialectical skills, broadcasts reputation, dispels doubts regarding things heard. . . Hence it is the discussion with men of the same branch of science, that is applauded by the wise."[29] The *Caraka-Saṁhitā* suggests *yukti* or heuristic reasoning as a valid and independent means of knowledge. Medical practitioners were advised to free themselves from bias and search for the truth dispassionately.[30]

According to Caraka, discussions can be friendly or hostile. In the Hindu tradition, even hostile discussions were an organized tradition in which scholars with differing opinions shared their point of view, and debated with each other.[31] The rule was to avoid any "celebration for the victor" or "any insult to the loser."[32] Since knowing truth was the purpose of a debate, it was the knowledge that became central and not the person. This discouraged a feeling of triumph or defeat for the participants. It was suggested that all assertions in a debate should be made in a polite manner. A person in anger can do anything to win, even inappropriate actions. Therefore, wise people debate in a polite manner.[33] This chapter of the *Caraka-Saṁhitā* reminds us of the Robert's Rules of Order that were established in the West centuries later, in the nineteenth century.[34]

2.3 RESPECT FOR KNOWLEDGE: THE ROLE OF A GURU

Hinduism does not enforce any undue restraint upon the freedom of human reasoning, the freedom of thought, or the will of an individual. Hindus' respect for knowledge is inherent in the core values of the religion. Respect for learning is obvious from the status that was bestowed on gurus. A festival, *guru-pūrṇimā*, is celebrated every year when Hindus offer thanks, love, and devotion to their gurus. This festival falls on the full moon day (*pūrṇimā*) in the month of *Āṣāḍha* (June - July). The *Śvetāśvatara-Upaniṣad* tells us that the Vedic knowledge is automatically revealed to a person who has the deepest love for God and the same love toward his teacher.[35]

[28] *Arthaśāstra*, 1: 2: 6–7.
[29] *Caraka-Saṁhitā*, *Vimānasthāna*, 8: 15.
[30] *Caraka-Saṁhitā*, *Sūtrasthānam*, 25: 32.
[31] *Caraka-Saṁhitā*, *Vimānasthāna*, 8: 16.
[32] *Caraka-Saṁhitā*, *Vimānasthāna*, 8: 17.
[33] *Caraka-Saṁhitā*, *Vimānasthāna*, 8: 22–23.
[34] Henry Martyn Robert (1837–1923) was an engineering officer in the U.S. Army. In 1875, he self-published a book, *The Pocket Manual of Rules of Order for Deliberative Assemblies*, in two parts. The book gained popularity and the first formal edition was published in 1876 with a new title: *Robert's Rules of Order*. The book is still a classic and frequently consulted by groups for smooth interactions and discussion.
[35] *Śvetāśvatara-Upaniṣad*, 6: 23.

In Hindu tradition, gurus[36] have a very high status that is comparable to parents and God. For example, Kabir, a famous medieval poet, shared a situation where his guru and God both appeared before him at the same time. In the Hindu tradition, dignitaries are greeted and honored by the host by touching their feet in the order of their comparative stature. Kabir's dilemma was who should he greet first, his guru or God? Kabir quickly resolved this dilemma as he realized that the guru should be the first one since he (guru) enabled him to see God.[37] Gurus shared their own personal experiences with disciples, guided disciples in their interactions with the community, taught social rules, taught various intellectual disciplines, and most important of all, became the spiritual guides of disciples. It was the guru who assigned a skilled trade or profession *varṇa*, later known as *jāti* (or caste), to a disciple after the completion of his or her education.

Hindus divided the human lifespan of 100 years into four stages (*āśrama*): *brahmacarya* (learning stage), *gṛhastha* (householder stage), *vānaprastha* (stage of teaching, and doing community service) and *sanyasa* (stage of contemplation and renunciation). *Brahmacarya* is the first stage of life where a child, after the infancy stage, joins *gurūkula* (*gurū* = teacher, *kula* = home or lineage), boarding school run by a guru in a natural environment (or forest). Parents left their children here in the care of the guru, to live in the school compound with the family of their guru and not with their parents to get appropriate education suited to their aptitude and inclinations. Irrespective of the social and economic status of the parents, each child had to live at the same standard of living as the guru's family. This allowed disciples to live near the guru, who could observe firsthand aptitudes and inclinations of disciples. The disciples understood the complexities of life quickly because of their continued association with the guru. Young children not only gained content-based knowledge from their guru, they also learned virtuous lifestyles and ethics. This system worked quite well for the ancient Hindus and they excelled in the sciences, along with other skill trades and disciplines.[38]

Gurus were considered to be authority figures, but were not considered infallible. It was considered a healthy practice for the disciples to raise questions about gurus' teachings to understand reality in their own way. The *Prasna-Upaniṣad* (Prasna = question), one of the principal *Upaniṣads*, is entirely based on the questioning between disciples and their teacher.

The role of the guru was and still is paramount to most Hindus. It does not stop after the *brahmacarya* stage; it continues for the rest of the life. Carl Jung, a noted psychoanalyst, emphasized the importance of the guru in the following words: ". . . practically everybody of a certain education, at least, has a guru, a spiritual leader who teaches you and you alone what you

[36] *Śikṣak, ācārya, śrotriya Upādhyāya, purohit* are other words for guru.

[37] *Guru Govind dāū khare, kāke lāgu pāya, balihārī guru āpnu jin Govind diyo batāya.* My teacher and God are in front of me. Whom should I prostrate first? It has to be the guru who taught me to recognize God.

[38] Joshi, 1972.

ought to know. Not everybody needs to know the same thing and this kind of knowledge can never be taught in the same way."[39]

Some of the famous teachers in Hindu history are Kauṭilya (fl. 300 BCE), teacher of Candragupta Maurya (reigned c. 321–297 BCE) and Vasiṣṭha (pre-historic), teacher of Lord Rāma. Kauṭilya laid the rules of administration for Candragupta, especially for psychological warfare, political philosophy, and economics, which are compiled in his book, *Arthaśāstra*. It was Kauṭilya's strategies that ended the Greek rule in a western state of India (now in Pakistan). Similarly, it was Vasiṣṭha who convinced King Daśaratha to allow Lord Rāma to relinquish the worldly comforts of his palace and live in a forest to protect sage Viśvāmitra's gurukula from *rākṣasa* (evil mongers).

2.4 *SMṚTI* (MEMORY), AN ANSWER TO BOOK BURNING

In the *gurukulas*, the ancient Hindus memorized their literature (mostly poetry) verbatim. The spoken words, not the written words, have been the basis of literary and scientific traditions of the Hindus. The people who memorized the texts were highly respected as they became the tools that could keep the tradition alive. This tradition continues even today. People who memorized *Vedas* or *Upaniṣads* are highly respected in today's Hindu society.[40] This memorization tradition was facilitated by composing their literature in Sanskrit either in stories or in poetry with a rhythm or pattern. Stories have characters with links with each other. Similarly, it is much easier to memorize a poem than a prose due to the rhythm. Sanskrit grammar was developed by Hindus to facilitate composition of poetry. Even math problems were composed in beautiful poetry in Sanskrit. The first written accounts in India are from the period of Aśoka (r. 269–232 BCE), the third emperor in line of the Mayura dynasty. He erected stone obelisks with his edicts inscribed in the stones rock edicts all over India. These edicts are useful guide to life in ancient India. However, such inscriptions or manuscripts are limited in number.

Yijing (also I-tsing, 635–713), a Chinese traveler who visited India, was impressed when he met people who could recite hundreds of thousands of verses of *Vedas*. "The *Vedas* have been handed down from mouth to mouth, not transcribed on paper or leaves. In every generation there exist some intelligent Brahmans who can recite 100,000 verses . . . This is far from being a myth, for I myself have met such men," writes Yijing.[41]

[39]Jung, Carl, *CW Letters*, 1973, vol. 1, p. 237. Jung (1875–1961) was an influential psychiatrist and thinker of the twentieth century. Hindu philosophy played an important role in his theories on symbolism and the unconscious mind.

[40]Vedi (or Bedi), Dvivedi, Trivedi, and Caturvedi are common last names (surnames) among brahmins, the people who were entrusted to preserve the *Vedas*. These names literally symbolize the number of *Vedas* memorized by them. *Dwi* means two and Dwivedi is the surname of a person who has memorized two *Vedas*. Similarly, *tri* and *catur* mean three and four, respectively. Thus, Trivedi and Caturvedi are the people who have memorized three or four *Vedas*, respectively. Of course, today these surnames are inherited from father to children without any connection to memorization.

[41]Takakusu, 1966, p. 182. I have thought of this statement along with the people who memorized *Vedas* in India. Suppose, it takes 10 seconds to narrate one verse. To narrate 100,000 verses, this equals to one million seconds. One million seconds equal a period of non-stop chanting for more than 11 days. How can one remember voluminous texts that take more than 11 days to read? Obviously, the person had to breathe in between, take rest, eat food, and sleep. This means that, in a rough

Al-Bīrūnī, an Islamic scholar who lived in India for some thirteen years during the eleventh century, wrote of the importance of poetic literature in popularizing science: "By composing their books in metres [poetry] they [Hindus] intend to facilitate their being learned by heart, and to prevent people in all questions of science ever recurring to a written text, save in case of bare necessity. For they think that the mind of man sympathizes with everything in which there is symmetry and order, and has an aversion to everything in which there is no order. Therefore, most Hindus are passionately fond of their verses, and always desirous of reciting them, even if they do not understand the meaning of words, and the audience will snap their fingers in token of joy and applause. They [Hindus] do not want prose compositions, although it is much easier to understand them."[42]

During the Islamic invasion of India, libraries were burnt. As an example, the libraries in Nalanda and Vikramsila were destroyed around 1200 CE by Bhkhtiyar Khilji. However, most of the sacred literature of the Hindus was easily reproduced because Hindus had memorized the poetic verses of their sacred literature. The fictional novel *Fahrenheit 451* by American writer Ray Bradbury dramatizes this feat. The title is based on the temperature at which paper catches fire. This novel deals with a futurist society where books were outlawed and firemen burned books. A small group of people countered this situation by memorizing books.

The tradition of memorization of sacred texts (or lack thereof) had a profound outcome for world cultures. Comparatively speaking, when the textual riches of Alexandria, China, Baghdad, and Rome were flamed, the glory of these cultures dissipated like smoke in the sky. In contrast, the Hindus could still salvage much of their textual riches.[43]

Memorization was facilitated by the abundant use of polysemous words in Sanskrit to maintain rhythm and tone. In some situations, the meanings of a word are so divergent that multiple interpretations can be made of the same sentence. It creates a dilemma. What is the original intended meaning of a verse? To make matter complex, the authors of the Hindu literature deliberately intended multiple meanings of their verses, depending on the expertise of a person. A layperson as well as a scholar could enjoy the hymns. As a result, scholars today argue with each other to validate their own interpretations. The problem becomes tense when a native interpretation is deemed biased by a foreign experts. In contrast to Catholicism, Judaism, and Islam, the contemporary Hindu literature is mostly produced by non-practitioners. For example, Max Müller, Monier-Williams, and Rudyard Kipling were devout Christians who translated the sacred Hindu texts with the intent of changing the religious landscape of India. Even after a century has elapsed, their translations are still popular in US, UK and Europe.

estimate, the person memorized a text that took about 25–30 days to narrate. Is it possible? Historical documents support such memorization.

[42]Sachau, 1964, vol. 1, p. 137.

[43]Montgomery and Kumar, 2000.

2.5 YOGA AND MEDITATION FOR SELF-IMPROVEMENT

Our physical body and mind are the basic instruments for all our actions, perceptions, and thoughts. It is the body and mind (popularly defined as two separate entities) that are the basic tools of our knowledge, virtues, and our happiness. In the modern culture, more emphasis is placed on the body. Gyms are popularly becoming a place where people go for exercise and sculpt their bodies for self improvement. Our icons are the supermodels and superathletes. What about the mind? Or, more importantly, a unison of body and mind?

The first organizing principle underlying human movements and postures is our existence in the gravitational field. The earth's gravitational field has an influence on every movement we make. The combination of the nervous system and skeletal muscles functioning together in gravity forms the basis of various yoga postures.[44] The practice of yoga perhaps is the most cost-effective and non-invasive treatment in many medical conditions. It is the most integrated science of self improvement where body and mind are both nourished that allows people to reach their fullest potential.

The Sanskrit term yoga comes from the root '*yuj*' which means "union," "join" or "balance." The term "union" defines the union of our physical self and the mind for some, while for others, a union of self with the divine. However, in both unions, a person transcends from everyday mundane existence to his/her fullest potential that leads to the understanding of the unity of all living creatures and ultimately to *mokṣa*—an ultimate goal for all Hindus. Body and mind are the vehicles for the Hindus that allowed them to liberate their soul from the cycle of birth and death. Body and mind are the Siamese twins that create synergy for this ultimate goal of liberation.

Yoga is all about balance; it is a balance of mind and body, a balance of strength and flexibility, a balance in a particular posture, even a balance in breathing from left and right nostrils that harmonizes the left and right brains. It is a philosophy; it is a way of life. It helps to avoid sickness. It helps a person to reach his/her best potential. Yoga is perhaps the oldest system of personal development that is effective.[45] Although the *Vedas* were the first to mention the importance of yoga, the *Upaniṣads* were the first to provide a systematic form of yoga. The first major treatise of yoga known to us is Patañjali's *Yoga-Sūtra*. The Indus seals of Harappa and Mohenjo-daro depicting human figurines in lotus postures in a meditative state provide archaeological support for the existence (Figure 2.1).[46]

In Patañjali's *yoga-sūtra*, yoga is defined as a system of eight limbs (*aṣṭāṅga-yoga*):[47] "Restraint (*yama*), observance (*niyama*), posture (*āsana*), breath-control (*prāṇāyāma*), sense-

[44] Coulter, 2001, 23.

[45] Coulter, 2001; Cowen, 2010; Eliade, 1969; Feuerstein, 1989; Iyengar, 1966; and Kulkarni, 1972. For a beginner who is interested in the medical aspects of yoga, Coulter's book is good. For the philosophical aspects, consult Eliade's book.

[46] Worthington, 1982, p. 9.

[47] Patañjali's *yoga-sūtras*, 1: 40.

Figure 2.1: A monk sitting in a lotus posture in a meditative state (taken from Wikimedia).

withdrawal (*pratyāhāra*), concentration (*dhāraṇā*), meditative-absorption (*dhyāna*) and enlightenment (*samādhi*) are the eight members [of Yoga]."[48]

Yoga postures (*āsana*) are an outcome of biomimicry practiced by the ancient Hindus. They observed various life forms—small and big, their life styles, the ways they exercised, the ways they cured themselves, the ways they relaxed, and the ways they avoided sickness. These studies evolved into a system of medicine called Ayurveda (science of life). It is no coincidence that many *āsana* are named after animals. The names of various postures are either based on their geometry or their similarity to an object, bird or animal. The following are some popular *āsana*: *dhanura-āsana* (bow posture), *garuḍa-āsana* (eagle posture), *krounc-āsana* (heron posture), *makar-āsana* (crocodile posture), *maṇdūka-āsana* (frog posture), *mayur-āsana* (peacock posture), *padma-āsana* (lotus posture), *trikoṇa-āsana* (triangle posture), *vakra-āsana* (curved posture), and *hala-āsanas* (plough posture).

Yoga is also a science that is cognizant of the bones, muscles, joints, organs, glands and nerves of the human body (biology). It uses the physics of balance in designing postures, the physics of motion and balance to allow changes in postures and a deep understanding of the strength and connectivity of various muscles to make the body strong and flexible. At the same

[48]Patañjali's *Yoga-Sūtra*, 2: 29.

time it is based on a deep understanding of the power and functioning of the mind (psychology) in controlling the thought processes (meditation) for optimum or holistic health.

The word yoga has appeared in the *Ṛgveda* to define the yoking, connection, achieving the impossible.[49] By yoga, one gains contentment, endurance of the pairs of opposites, and tranquility, tells the *Maitreyī-Upaniṣad*.[50] "When cease the five senses, together with the mind, and the thoughts do not stir. That, they say, is the highest course. This they consider as yoga, firm holding back of the senses. Then one becomes undistracted. Yoga, truly, is the origin and end," suggests the *Kathā-Upaniṣad*.[51] In other words, yoga is alpha and omega of the art of self-improvement.

Here "origin and end" means that yoga is essentially involved in knowledge and experience; it is a process at all stages. Yoga is advocated for the knowledge or realization of the self (*ātman*).[52] Patañjali equated yoga with *samādhi* (tranquil state) in the very first verse of his book. Yoga is not a mere abstract speculation of human mind; it is real with a concrete referent: "[This supreme ecstasy] is near to [him who is] extremely vehement [in his practice of Yoga]."[53]

Our body and mind are intimately linked. If the muscles of our body are toned and relaxed, it is easier for our mind to relax. Similarly, if our mind is anxious, it directs stimuli to our physical body and changes the chemical composition and physical state of our muscles. The outcome is the drainage of physical and emotional energies. The ancient Hindus realized the intimate link of mind and body, and formulated exercises for both. The physical body received its vital energies through yogic postures (*āsana*), while mind gained its vital energies through meditation.

An active organ receives a larger flow of blood than an inactive organ. Blood is an essential ingredient for the proper functioning of the various organs of body, and these organs get enriched due to a higher and more efficient transfusion of oxygen through our lungs. The yoga postures work on the body frame as well as on the internal organs, glands, and nerves. Most joints and organs are put into isometric or other ranges of motions.

Most people take short and shallow breaths throughout the day using their chest. This kind of breathing does not allow our "lungs to expand and soak up the oxygen."[54] Although hyperventilation or heavy breathing can be useful in the short term by boosting sympathetic nervous system activity, a better way to breathe is using the abdomen and diaphragm, called the belly breath and take slow and steady breaths. You need to use your diaphragm, which is the muscle underneath your lungs. When the diaphragm flexes, it pulls down and opens the lower lobes of your lungs, allowing more air inside. Chest breathing comes from stress; it is a reaction to stress.

[49] *Ṛgveda*, 1: 34: 9; 3: 27: 11; 7: 67: 8; 10: 114: 9; Dasgupta, 1963, vol. 1, p. 226.

[50] *Maitreyī-Upaniṣad*, 6: 29.

[51] *Kathā-Upaniṣad*, 6: 10–11.

[52] *Kathā-Upaniṣad*, 2: 12; *Muṇḍaka-Upaniṣad*, 3: 2: 6; *Śvetāśvatara-Upaniṣad*, 1: 3: 6–13; *Maitreyī-Upaniṣad*, 6: 18, 19, 27.

[53] Patañjali's *Yoga-Sūtra*, 1: 21.

[54] Dollemore, Giuliucci, Haigh, Kirchheimer and Callahan, 1995, p. 152; Gilbert, 1999a and 1999b.

All yoga exercises start with basic deep breathing techniques along with proper *āsanas*. It serves two purposes: first, to bring an optimum amount of oxygen into the lungs; second, to control the mind by controlling the breath. Deep breathing is not the fast pumping action of our lungs where we take a fast deep breath and puff it out; it involves a controlled rhythmic action to fill the lungs with air, to retain it for a brief period, and slowly exhale. The whole process has four parts: inhalation (*pūraka*), retention (*kumbhaka*), exhalation (*recaka*), and suspension (*kumbhaka*).

The word meditation derives from a Latin word, *mederi*, which means to heal. It is the healing of a mental affliction caused by psychological stress. Managing our thought processes is a key to managing the stress that affects our lives. Meditation can help a person find new ideas and practical answers to problems. It gives ample stillness to the mind to think properly in order to make proper judgments. It helps the mind to control emotion without suppression but with an outlet where the emotional waste could be discarded in order to become more at peace with the world. For this reason, the *Maitreyī-Upaniṣad* considers meditation as essential to realize God, along with knowledge (*vidyā*) and austerity (*tapas*).[55]

Meditation is a process of *knowing* for the Hindus. The Sanskrit words that reflect meditation are: *cintan*, *dhyāna*, or *manana*. Nowhere in the world was the art of meditation as perfected as was done by the ancient Hindus. The *Śvetāsvetara-Upaniṣad* tells us that meditation and yoga are the way to know the self-power (*āṭma-śakti*) of God within us.[56] The *Chāndogya-Upaniṣad* tells us that all people who achieved greatness in the past did so with the help of meditation.[57] The practice of concentration (*ekāgratā* or *dhārṇa*) is the endeavor to control the two generative sources of mental fluidity: sensory activity (*indriya*) and subconscious activity (*saṃskāra*).[58] It is difficult to achieve pursuit of one object (*ekāgratā*, single-mindedness) with a tired body or restless mind, and with unregulated breathing.

Meditation can decrease our reaction time, increase alertness and improve the efficiency of a person. Insomnia, headache, lack of appetite, shaking of the hands and other symptoms can be either reduced or nearly eliminated. It also helps in asthma, anxiety, high blood pressure, back pain, heart disease, etc. Concentration of the mind acts like a focusing lens (convex lens) of Sun's rays. It concentrates thought and invokes miraculous powers to the mind's activities. The only way to understand the impact of yoga is to go through the experience.

The diaphragmatic movements provide a "massaging action to the heart" as well as to the *inferior vena cava* as the latter passes through the diaphragm, thus propelling the blood forward toward the heart and can be labeled as the "second heart."[59] In the Universidade Federal de São Paulo, Brazil, Danucalov *et al.* investigated the changes in cardiorespiratory and metabolic intensity resulting from *prāṇāyāma* and meditation during the same *hatha* yoga session. Nine

[55] *Maitreyī-Upaniṣad*, 4: 4.
[56] *Śvetāsvetara-Upaniṣad*, 1: 3.
[57] *Chāndogya-Upaniṣad*, 7: 6: 1.
[58] Eliade, 1975, p. 62.
[59] Thomas, 1993; taken from Gilbert, 1999a

yoga instructors were subjected to analysis of the air exhaled during the three periods, each of 30 minute duration: rest, respiratory exercises, and meditation. The oxygen uptake and carbon dioxide output were proven to be statistically different during the active sessions (*prāṇāyāma* and meditation) compared to the rest phase. In addition, the heart rate showed decreased levels during rest as compared to meditation. Therefore, the results from this study suggest that meditation reduces the metabolic rate while the *prāṇāyāma* techniques increases it.[60]

Researches have shown that yoga can help people suffering from asthma, heart disease, high blood pressure, type 2 diabetes, and obsessive-compulsive disorder, lower their dosage of medications, and sometimes eliminate the use of medication.[61] Practicing yoga not only relieves temporary symptoms such as headaches, sinus pressure and hot flashes, but it can also improve health during more serious medical conditions such as cancer, diabetes, anxiety, and heart disease, to name a few. "Yoga is not a panacea, but it is powerful medicine indeed for body, mind, and spirit," suggests Dr. McCall, a medical doctor who studied and practiced the effects of yoga.[62]

[60] Danucalov *et al.*, 2008.
[61] McCall, 2007, p. 43.
[62] McCall, 2007, p. XIX.

CHAPTER 3

The Hindu Mathematics

"Of the development of Hindu mathematics we know but little. A few manuscripts bear testimony that the Indians had climbed to a lofty height, but their path of ascent is no longer traceable," wrote Cajori in his book, *A History of Mathematics*.[1] This indicates the status of scholarship on Hindu mathematics in 1893 when Cajori first published this book. However, the scholarship of the last hundred years has filled many gaps in our understanding of Hindu mathematics.

Most mathematics textbooks are devoid of historical anecdotes and stories. Mathematics is taught as a discipline that someday somehow appeared in fully developed form. Therefore, students of mathematics do not learn much about the dynamic aspect of the discipline. This led George Sarton, a leading historian of science and chemistry in the twentieth-century, a former professor at Harvard University, to write, "if the history of science is a secret history, then the history of mathematics is a doubly secret, as secret within a secret."[2] Mathematics is also the discipline where the non-Western cultures have made significant contributions. Mathematics owes a lot to its Hindu and Middle Eastern roots.[3]

3.1 THE HINDU NUMERALS

The number system that we use today in most of the civilized world has come to us from the ancient Hindus. This system of counting is so simple that it is difficult to realize its profundity and importance.[4] Young children all over the world are generally taught this counting system first and the alphabet of their native language later. Children are taught to write eleven as one and one (11), written side-by-side, which they learn without much difficulty. "Our civilization uses it unthinkably, so to speak, and as a result we tend to be unaware of its merits. But no one who considers the history of numerical notations can fail to be struck by the ingenuity of our system, because its use of the zero concept and the place-value principle gives it an enormous advantage over most of the other systems that have been devised through the centuries."[5]

[1]Cajori, 1980, p. 83.

[2]Taken from Dauben, Joseph W., Mathematics: a Historian's Perspective, a chapter in the book by Chikara, Mitsuo and Dauben, 1994, p. 1.

[3]Bag, 1979; Colebrooke, 1817; Datta and Singh, 1938; Ifrah, 1985; Joseph, 1991; Rashed, 1996; Smith, 1925; Srinivasiengar, 1967.

[4]Ifrah, 1985, p. 428. Ifrah's book is still one of the most engaging and scholarly book on numerals in various ethnic cultures. The other noted books are by Calinger, 1999; Cook, 1997; Katz, 1993; Smith, 1925; Smith and Karpinski, 1911; Suzuki, 2002.

[5]Ifrah, 1985, p. 428.

Several civilizations such as the Egyptians, Chinese, and Romans used the additive decimal systems although their symbols were different. In their system, multiple repetition of symbols were used and added to increase magnitude. Thus, in the Roman system, X means 10 and XX becomes (10 + 10) = 20. Also, XVI amounts to 10 + 5 + 1 = 16. The Romans also used subtractive symbolism where IX and IL represents 9 (as 10–1) and 49 (as 50–1), respectively.

3.1.1 THE WORD-NUMERALS

In an oral tradition, numbers are chosen as words that can be used in a narrative. This happened with the ancient Hindus where words representing numbers were chosen in verses of the sacred literature. The word-numerals in Sanskrit language are written similar to the way numbers are written in German language. For example, 23 is called three and twenty (*drei und zwanzig*) in German while twenty and three in the English system. Similarly, fourteen is called *vier-zehn* (four and ten) and fifty-three is called *drei und fünfzig* (three and fifty). In the Sanskrit language, the mother language of all Indo-European languages, the language of the *Vedas*, twelve is written as two and ten,[6] 34 as four-thirty,[7] 53 as three and fifty,[8] 77 as seven and seventy.[9] just like the German system.

For compound numerals, the number of higher order was placed as qualifier and the lower as qualified. For example, eleven is defined as ten qualified by the addition of one, thus giving *eka-dasa*, translated as one-ten. The number 720 is denoted as seven-hundred-twenty[10] and the number 60,099 is written as sixty-thousand-nine-ninety[11] in the *Ṛgveda*. The *Ṛgveda* mentions "three thousand and three hundred and nine-thirty (3339)"[12] as the count of people in a *yajna*, a holy gathering where worshipping is done around fire. The *Atharvaveda* defines a hundred, thousand, myriad, hundred-million.[13] Similarly, consistent with the practice in the *Ṛgveda*, the *Baudhāyana-śulbasūtras* expressed the number 225 as (200 and 5 and 20)[14] and the number 187 as (100 + 7 + 80).[15]

For the numbers provided in the *Vedas* and *Śulbasūtras*, the following points are clearly established:

1. The numbers from one to ten have specific names.

2. After ten, specific words are designated to 20, 30, . . . 100. All other numbers in between are defined as a combination of these words.

[6] *Ṛgveda*, 1: 25: 8.
[7] *Ṛgveda*, 1: 162: 18; *Ṛgveda*, 10: 55: 3.
[8] *Ṛgveda*, 10: 34: 8.
[9] *Ṛgveda*, 10: 93: 15.
[10] *Ṛgveda*, 1: 164: 11.
[11] *Ṛgveda*, 1: 53: 9.
[12] *Ṛgveda*, 10: 52: 6.
[13] *Atharvaveda*, 8: 8: 7.
[14] *Baudhāyana-śulbasūtras*, 16: 8.
[15] *Baudhāyana-śulbasūtras*, 11: 2.

3. After 100, new words are assigned to 1,000, 10,000, 100,000, etc.

4. In word numerals, numbers between 10 and 20 were defined as a number above ten and then 10. For example, 12 is defined as 2 and 10 (*dvi-daśa*). A similar practice was made for other numbers. For example, 99 is written as (9 + 90), 76 as (6 + 70), etc.

The ancient Hindus wondered about the total number of atoms in the universe, the age of the universe, size of atom, etc. To define these physical quantities, they used really large numbers as well as really small numbers. Al-Bīrūnī, an Islamic philosopher who lived in India during the eleventh-century, criticized the Hindus for their passion for large numbers: "I have studied the names of the orders of the numbers in various languages with all kinds of people with whom I have been in contact, and have found that no nation goes beyond thousand. The Arabs, too, stop with the thousand, which is certainly the most correct and the most natural thing to do. Those, however, who go beyond the thousand in their numeral system are the Hindus, at least in their arithmetical technical terms."[16] This criticism of al-Bīrūnī demonstrates that the ancient Hindus were simply much ahead of their time. I am fairly confident that if al-Bīrūnī were to write his book today, he would not criticize the Hindus for their fondness of large numbers. Al-Bīrūnī mentioned 10^{19} as the largest number used by the Hindus.

3.1.2 THE PLACE-VALUE NOTATIONS

The Greeks, Egyptians, and Romans, did not use place-value notations in writing numbers while the Babylonians, Chinese, Mayans, and Hindu, did use. The Babylonian system was base-sixty while the Mayan system was base-twenty. The current place-value system (also called position-value system) that is base-10 is Hindu in origin. This is certain. The uniqueness of the Hindu system lies in the fact that the position of a number qualifies its magnitude. Tens, hundreds, or thousands were not represented by different signs; they are represented by using digits in different positions. For example, the one is in second place in 10 (ten), in third place in 100 (hundred), and in fourth place in 1,000 (thousand).

In a positional- or place-value system, a number, represented as $x_4x_3x_2x_1$ can be constructed as follows:

$$x_1 + (x_2 \times 10^1) + (x_3 \times 10^2) + (x_4 \times 10^3)$$

Where x_1, x_2, x_3, and x_4 are nonnegative integers that have magnitudes less than the chosen base (ten in our case). As you may have noticed, the magnitude of a number increases from right to left. For example, the number 1234 will be written as

$$4 + (3 \times 10^1) + (2 \times 10^2) + (1 \times 10^3)$$

Similarly,

$$1.2345 = 1 + \frac{2}{10} + \frac{3}{10^2} + \frac{4}{10^3} + \frac{5}{10^4}$$

[16]Sachau, 1964, vol. 1, p. 174.

In India, Āryabhaṭa I (born 476 CE) used a positional value system in describing numbers, did not even bother to explain much about it, and claimed it to be ancient knowledge. This indicates that the system was prevalent then.[17]

The earliest written record of the place-value notation comes from Vāsumitra, a leading figure of Kaniska's Great Council. According to Xuan Zang (also known as Hiuen Tsang, 602–664), Kusāna King Kaniska (144–178 CE) called a convocation of scholars to write a book, *Mahāvibhāsa*. Four main scholars under the chief monk, named Pārśva, wrote the book in 12 years. Vāsumitra was one of the four scholars. In this book, Vāsumitra tried to explain that matter is continually changing as it is defined by an instant (time), shape, mass, etc. As time is continually changing, therefore, matter is different in each situation although its appearance and mass do not change. He used an analogy of the place-value notation to emphasize his point. Just as location of digit one (1) in the place of hundred is called hundred (100) and in place of thousand (1,000) is called thousand, similarly matter changes its state (*avasthā*) in different time designations.[18]

New explanations are generally given in terms of known and established facts. Thus, the very reason Vāsumitra used place-value notation as an example to define change in matter establishes that the place-value notation was considered as an established knowledge during the early Christian era.

In modern perspective, just imagine reading the values of various stocks in a newspaper. In a quick scan, you can recognize easily that 1089 is greater than 951. All you need to see is that the first number has four digits while the second number has only three. This is enough for a quick comparison. In contrast, in the Roman numerals, XC (90) is five times more in magnitude than XVIII (18). This is not easy to figure out in a quick glance. Also, mathematical operations of multiplication, division, addition and subtraction become much simpler in a place-value notation.

3.2 FROM *ŚŪNYATĀ* AND *NETI-NETI* TO ZERO AND INFINITY (*ANANTA*)

Zero and infinity are perhaps two grandest concepts ever invented in mathematics. Zero is one of the top hundred discoveries/inventions ever produced in history, as listed by American Association for the Advancement of Science in its flagship journal, *Science*, in year 2000. It is the last numeral invented by the Hindus that prompted the natural philosophers to dump the abacus. It became much easier to perform calculations on a tablet or paper. "In the simple expedient of

[17] *Āryabhaṭīya*, Gola, 49–50.
[18] Ruegg, 1993; Sinha, 1983, p. 130.

cipher [zero], which was permanently introduced by the Hindus, mathematics received one of the most powerful impulses," writes Cajori in his book *A History of Mathematics*.[19]

Zero is a numeral and it has the same status as any other numerals, all in the absence of any magnitude. The presence of zero indicates a specific absence of the symbols of 1, 2, . . . , 9 at that location. Zero is thus a sign that defines no value or a missing value in a particular location in a number. It also defines the starting point in measurements, such as the coordinate axes, meter sticks, stop watches, and thermometers. Zero is the denial of number in a particular location and gains its meaning from digits that are on the left of it. Zero plays the role of a number and at the same time signifies the metaphysical reality of the absence of substance (emptiness). In Hindu philosophy, the terms *śunyata* and *neti-neti* define emptiness which evolved into a mathematical reality in the form of zero, as explained later.

Zero, as a "void," is an integral part of the Hindu philosophy. The *nirguṇa-rūpa* (nonmanifested-form, *amūrta-rūpa*) of God is worshiped by the Hindus. In this form, no attributes can be assigned to God. Nothingness (*śūnyatā*, emptiness or void) as such, in Hindu tradition, has a substance; it is not an absence of everything with 100% mathematical certitude; it is an absence of all attributes that are within the realm of *māyā* (illusion of the manifested world).

In the *nirguṇa* manifestation, God is beyond any attributes yet the source of all. He is nowhere and yet everywhere. The mathematical symbol zero has similar qualities. Zero has no magnitude and, therefore, is present in every number as $a + 0 = a$. It shows its presence when associated with a number in the decimal system.

The concept of *neti-neti* (not this, not that; essential for *nirguṇa-rūpa*, as we cannot assign any particular attribute to God) dates back to the *Vedic* and *Upaniṣadic* periods. The *Bṛhadāraṇyaka-Upaniṣad* explains the Supreme God (ultimate reality) by defining God as the absence of all the attributes (*neti-neti*).[20]

The absence of a number in place-value notation has a meaningful function. It does not have a similar usefulness in any other system that is not place-value. For example, the location of two digits 5 and 4 can be fifty four (54), five hundred four (504), five hundred forty (540), etc., depending on the location of the two digits. It is only natural to assign a symbol—a circle or a dot—for this absence of a number, for convenience. This is how zero became so central to the place value system and mathematics.

The concept of zero was known in ancient India philosophically during the *Upaniṣad*-period. We do know that zero was used as mathematical entity in *Chandāḥ-sūtra*. When did it become, along with a philosophical entity, to mathematical reality is not clearly established. In a short aphorism, to find the number of arrangements of long and short syllables in a meter containing *n* syllables, zero is defined: "[Place] two when halved, when unity is subtracted then

[19]Cajori 1980, p. 147; Accounts on the invention of zero are provided by Bronkhorst, 1994; Datta, 1926; Gupta, 1995; Kak, 1989, 1990; Ruegg, 1978. Ruegg has provided perhaps the best review that includes philosophical insights and historical developments.

[20]*Bṛhadāraṇyaka-Upaniṣad*, 3: 9: 26.

(place) zero . . . multiply by two when zero . . ."[21] Piṅgala was the younger brother of eminent grammarian Pāṇini, who lived near Peshawar around 2850 BCE[22]

King Devendravarman of Kalinga, Orissa inscribed his deed on a copper plate in 681 CE. This deed provides an archaeological evidence of place-value notation. It lists twenty as two and zero (20) in a place-value notation.[23] In the Chaturbhuj Temple, Gwalior Fort, in Gwalior, India, a plaque is mounted where 270 number is listed with zero as a symbol. The plaque is about 1500 years old. The *Bakhshālī manuscript* mentions *śūnya* for zero at hundreds of places in the text.[24] By the seventh-century, this concept already reached in the Far-East, where inscriptions of zero, in writing 605, is carved on a sandstone. This stone was discovered at the archaeological site of Trapang Prei, in Kratie province, in northeastern Cambodia.

By the eighth-century, the concept of zero was already in China. Zero is mentioned and symbolized as a dot in Chapter 104 of the *Kaiyun Zhanjing* (*Astronomico-astrological Canon*), a book written during the reign of Kaiyun (713–741 CE): "Each individual figure is written in one piece. After the nine, the ten is written in the next row. A *dot is always written in each empty row* and all places are occupied so that it is impossible to make a mistake and the calculations are simplified."[25] The dot is an obvious reference to zero. This book was written by Qutan Xida (Gautama Siddārtha), an Indian scholar who settled in China, between 718 and 729 CE during the Tang dynasty.[26] This book also contains Āryabhaṭa's sine tables.

When the Arabs learned about zero, they literally translated the Sanskrit word *śūnya* (empty) into *sifr* (empty) in Arabic. "While the Arabs, as we have learned, did not invent the cipher [zero], they nevertheless introduced it with the Arabic numerals into Europe and taught Westerners the employment of this most convenient convention, thus facilitating the use of arithmetic in everyday life . . . al-Khwārizmī . . . was the first exponent of the use of numerals, including the zero, in preference to letters. These numerals he called Hindu, indicating the Indian origin."[27] Leonardo Fibonacci (1170–1250) called zero as *zephir*, in Latin, in his book, *Liber Abaci*.[28] Adelard of Bath (1080–1152) used the term *cifrae* in his translation of al-Khwārizmī's

[21] Piṅgala's *Chandāḥ-sūtra*, 7: 28, 29, 30; Datta and Singh, 1938, vol. 1, p. 75. This quotation does not indicate the origin of number zero but provides a testimony that the zero was used as a mathematical number in India. Similar accounts of zero are also made elsewhere in the same manuscript (*Chandāḥ-sūtra*, 3: 2 and 17; 4: 8, 11, and 12; 18: 35, 44, 48 and 51).

[22] Shyam Lal Singh, Piṅgala Binary Number, in the book by Yadav and Mohan, 2011, p. 121.

[23] Filliozat, 1993.

[24] Hayasi, 1995, p. 210, 213. The Bakhshālī manuscript consists of seventy fragmentary leaves of birch bark and is presently preserved in the Bodleian Library at Oxford University. The original size of a leaf is estimated to be about 17 cm wide and 13.5 cm high, containing mathematical writings. The manuscript was accidentally found in 1881 near a village, Bakhshālī, that is now near Peshawar in Pakistan. It is currently preserved in Bodleian Library at Oxford University. In his detailed analysis, Hayasi assigns the seventh century CE as the date when this manuscript was written. However, recently researchers at the Bodelian library at Oxford University investigated an old copy of Bakhshālī manuscript and found it to be written during the third or fourth century using carbon-dating—some five hundred earlier than previously thought. For more information, http://www.bodleian.ox.ac.uk/news/2017/sep-14.

[25] Martzloff, 1997, p. 207

[26] Yoke, 1985, p. 83

[27] Hitti 1963, p. 573.

[28] Horadam, 1975; Sigler, 2002, p. 17.

Zīj al-Sindhind.[29] The word zero in the English language evolved from the terms in Latin and Italian.

Opposite to the *nirguṇa-rūpa* approach to define God, the ancient Hindus tried to assign attributes connected to God. Once you start the process of assigning attributes, there is no limit; you can continue for ever. This led to the concept of infinity. For this reason, in the ancient literature of the Hindus, God is also named as *ananta* (infinity). The *Bṛhadāraṇyaka Upaniṣad* (5: 1) tells us: "The world there is full; The world here is full; Fullness from fullness proceeds. After taking fully from the full, It still remains completely full." When you take away something from a given quantity, it becomes less. Simple arithmetic dictates this. However, it all fails when we deal with infinity. You subtract something from infinity, it still remains infinity. So when you take away "fully from the full," the remaining is still the same, infinity. In *Surya Prajnapati*, a text that was written around 400 BCE, numbers are defined as enumerable, innumerable, and infinite. This text defined infinity in one direction, infinity in two directions (infinity in area), infinity in three directions (infinity in volume), and perpetually infinite.

3.3 THE BINARY NUMBER SYSTEM

A system in which numbers are represented as linear combinations of powers of two (2) is called the binary number system. This is a positional-numeral system employing only two kinds of binary digits, namely 0 and 1. The importance of this system lies in the convenience of representing decimal numbers using a two-state system in computer technology. A simple "on-off" or "open-closed" system can effectively represent a number. Similarly, a set of condensers "charged" or "not-charged" can represent a number, or a set of two different voltages on a device can effectively represent a number. For this reason, the binary number system is popular in electronic circuitry.

The presence of a binary number system is an example of the place-value notational system. In this system, the number 3 can be written as $(1 \times 2^1 + 1)$. In the binary system, we write 3 as 11. To understand the system, a few examples are provided in Table 3.1.

Using this system, a somewhat larger number 444 is represented as 110111100 $[(1 \times 2^8) + (1 \times 2^7) + (0 \times 2^6) + (1 \times 2^5) + (1 \times 2^4) + (1 \times 2^3) + (1 \times 2^2) + (0 \times 2^1) + 0]$.

To convert a number into a binary number, divide the number by two. If there is one as remainder, then write one (1). If it is divisible, then write zero. For example, to write 45 in the binary system, let us divide by two:

$45 \div 2$. The resultant is 22 and the remainder is 1. Let us write 1. $22 \div 2$. The resultant is 11 and the remainder is 0. Let us write 0. $11 \div 2$. The resultant is 5 and the remainder is 1. Let us write 1. $5 \div 2$. The resultant is 2 and the remainder is 1. Let us write 1. $2 \div 2$. The resultant is 1 and the remainder is 0. Let us write 0. $2 \div 1$. The resultant is 0 and the remainder is 1. Let us write 1.

[29]Neugebauer, 1962, p. 18.

Table 3.1: Decimal Numbers and Their Binary Equivalent

Decimal Numbers	Binary Equivalent
$1 = (0 \times 2^1) + 1$	01
$2 = (1 \times 2^1) + 0$	10
$3 = (1 \times 2^1) + 1$	11
$4 = (1 \times 2^2) + (0 \times 2^1) + 0$	100
$5 = (1 \times 2^2) + (0 \times 2^1) + 1$	101
$6 = (1 \times 2^2) + (1 \times 2^1) + 0$	110
$7 = (1 \times 2^2) + (0 \times 2^1) + 1$	111
$8 = (1 \times 2^3) + (0 \times 2^2) + (0 \times 2^1) + 0$	1000
$9 = (1 \times 2^3) + (0 \times 2^2) + (0 \times 2^1) + 1$	1001
$10 = (1 \times 2^3) + (0 \times 2^2) + (1 \times 2^1) + 0$	1010
$11 = (1 \times 2^3) + (0 \times 2^2) + (1 \times 2^1) + 1$	1011
$15 = (1 \times 2^3) + (0 \times 2^2) + (1 \times 2^1) + 1$	1111
$17 = (1 \times 2^4) + (0 \times 2^3) + (0 \times 2^2) + (0 \times 2^1)+ 1$	10001

By taking the first quotient 1 and the remainders in the reverse order, we can write 45 in binary number: 101101. This number implies $(1 \times 2^0) + (0 \times 2^1) + (1 \times 2^2) + (1 \times 2^3) + (0 \times 2^4) + (1 \times 2^5)$. This converts to (1 + 4 + 8 + 32) and equals to 45. Interestingly, this rule is provided in *Piṅgala-Sūtra* (8: 24–25) in a cryptic language which is nicely explained by Barend A. van Nooten using history documents that provided commentary to Piṅgala's verse.[30] Recently, Shyam Lal Singh has explained the rules of poetic metrics in an article, Piṅgala Binary Numbers, in the book, *Ancient Indian Leaps into Mathematics*.[31]

The ancient Hindus carefully defined the methodology to write hymns which involved the study of language and prosody. This allowed a verse to have specific rhythm (metrical structure) in chanting. The first comprehensive treatise that is known to us was written by Piṅgala, called *Chandāḥ-sūtra* (or Candāh) or *Chandāḥ-śāstra*.[32]

In Sanskrit, the science of versification (prosody) consists of verse feet, *padas*, that are composed of syllables: light (*laghu*) and heavy (*guru*). A knowledge of metrics is essential which is determined by the permutations and combinations of short and long syllables. Each verse usually consists of a set of four quarter verses, called *pāda*s. All verses contain the same number of syllables. For example, most verses in the Sanskrit language are either 8-, 11-, or 12-syllabic.

[30]van Nootan, 1993. Barend A. van Nooten is a former professor of Sanskrit at the University of California at Berkeley. He is known for his co-authored books, *Rigveda: A Metrically Restored Text* and *The Sanskrit Epics*.

[31]Yadav and Mohan [Editors], 2011.

[32]Weber, 1863; taken from van Nooten, 1993, also reproduced in Rao and Kak, 1998.

These quarters, (*pādas*), which are again subdivided into various groups or subgroups, depending on the number syllables (*akṣara*) in each quarter and the placement of short and long syllables. The vowels, such as a, i, u, ṛ, and ḷ are short syllables while ā, e, ai, ī, o, au, and ū are long syllables. There are some more rules to define the short and long syllables which are beyond the scope of this book.[33]

Piṅgala gave the following rule: "(Place) two when halved" "when unity is subtracted then (place) zero;" "multiply by two when zero;" "square when halved."[34]. It was van Nootan's expertise in Sanskrit metrics that allowed him to discover this unique binary system. In this system, each syllable is assigned a numerical value, based on its position in the meter. The work of Piṅgala in *Candāḥ-sūtra* definitely shows that he knew the place-value system of numeric notations and used a binary numerical base, and not base-10.[35]

The discovery of binary numbers is generally attributed to Gottfried Leibniz (1646–1716 CE) at the end of 17th century. Leibniz is said to have come up with the idea when he interpreted Chinese hexagram depictions of Fu Hsi in *I-Ching* (*The Book of Changes*) in terms of a binary code.[36] Piṅgala, in van Nooten's views, did not provide the further applications of the discovery. However, this knowledge was available to Sanskrit scholars of meterics.[37] "Unlike the case of the great linguistic discoveries of the Indians which directly influenced and inspired Western linguistics, this discovery of the theory of binary numbers has so far gone unrecorded in the annals of the West," remarks van Nooten.

3.4 THE FIBONACCI SEQUENCE

Leonardo Fibonacci (1170–1250) was an Italian mathematician who popularized Hindu numerals in the Western world by writing a book, *Liber Abaci*. This book played an important role in the growth of mathematics in Europe. Fibonacci came in contact with Hindu mathematics during his stay in Bugia, located on the Barbary Coast of Africa. In Fibonacci's own account, his father was a public official there to help the visiting Pisan merchants there. His father wanted Fibonacci to learn mathematics for "a useful and comfortable future." He arranged some lessons in mathematics for the young Fibonacci from well known scholars. This is how Leonardo learned about Hindu numerals. Leonardo recognized the superiority of this new system over his native Roman numeral system and wrote the above-mentioned book about it.

The Fibonacci sequence is connected with cumulative growth, and plays a role in various number games and natural phenomena, including a botanical phenomenon called phyllotaxis

[33]For more information, read Singh, Shyam Lal, Piṅgala Binary Numbers, in the book, *Ancient Indian Leaps into Mathematics*. Van Nooten, 1993 and Datta and Singh, 1962 are the other resources to understand this system.

[34]Datta and Singh, 1962, vol. 1, p. 76; van Nooten has provided a slightly different translation. However, both systems provide similar results.

[35]van Nooten, 1993

[36]Loosen and Vonessen, 1968, p. 126–131; reference taken from van Nooten, 1993.

[37]van Nooten, 1993.

where the arrangement of leaves on a system is studied.[38] The seeds-pattern in a sunflower follow the Fibonacci sequence.

The following is the problem that Fibonacci posed in *Liber Abaci* that is well known today as the Fibonacci sequence: "A certain man had one pair of rabbits together in a certain enclosed place, and one wishes to know how many are created from the pair in one year when it is the nature of them in a single month to bear another pair, and in the second month those born to bear also. Because the above written pair in the first month bore, you will double it; there will be two pairs in one month. One of these, namely the first, bears in the second month, and thus there are in the second month 3 pairs; of these in one month 2 are pregnant, and in the third month 2 pairs of rabbits are born, and thus there are five pairs in the month; in this month 3 pairs are pregnant, and in the fourth month there are 8 pairs, of which 5 pairs bear another 5 pairs; these are added to the 8 pairs making 13 pairs in the fifth month; these 5 pairs that are born in this month do not mate in this month, but another 8 pairs are pregnant, and thus there are in the sixth month 21 pairs; to these are added the 13 pairs that are born in the seventh month; there will be 34 pairs in this month; to this are added the 21 pairs that are born in the eighth month; there will be 55 pairs in this month; to these are added the 34 pairs that are born in the ninth month; there will be 89 pairs in this month; to these are added again the 55 pairs that are born in the tenth month; there will be 144 pairs in this month; to these are added again the 89 pairs that are born in the eleventh month; there will be 233 pairs in this month. To these are still added the 144 pairs that are born in the last month; there will be 377 pairs, and this many pairs are produced from the above written pair in the mentioned place at the end of the one year."[39]

Based on the assumptions, at the beginning of the second month, there will be two pairs. After the second month, there will be three pairs. In the third month there will be 5 pairs. In the consecutive month 8, 13, 21, 34, 55, 89, 144, 233, and 377 pairs will be there. Thus, 1, 2, 3, 5, 8, 13, 21, 34, 55, 89, 144, 233, and 377 pairs of rabbits will be available. As one can notice, in this number sequence,

$$x_n = x_{n-1} + x_{n-2}$$

Also,

$$x_{n+1} \times x_{n-1} = x_n^2 + (-1)^n$$

It was the French mathematician Édouard Lucas (1842–1891) who assigned this mathematical series as "the Fibonacci sequence." Later, an Oxford botanist, A. H. Church, recognized that the number of seeds in the spiral patter of sunflower head match with the Fibonacci sequence numbers. In 1963, an International Fibonacci Society was formed to study related topics.[40]

In India, this sequence appeared in the science of hymn-composing, or metrics, just like the binary numbers that were discussed in the previous section. In one particular category, the

[38] For more information, see Hoggatt, 1969.

[39] Sigler, 2002, p. 404–405.

[40] Gies and Gies, 1969, p. 81–83.

number of variations of meters having 1, 2, 3, . . . *morae* (syllabic instant) are 1, 2, 3, 5, 8, 13, . . ., respectively, the so-called Fibonacci numbers. Ācārya Virāhanka (lived sometime between 600–800 CE), Gopāla (before 1135 CE) and Hemachandra (ca. 1150 CE) had provided this sequence before it was suggested by Leonardo Fibonacci. To understand this metrical science, a knowledge of Sanskrit is required. Readers can get more information on this from the work of Permanand Singh that is published in a prestigious journal, *Historia Mathematica*. He concludes that "the concept of the sequence of these numbers in India is at least as old as the origin of the metrical sciences of Sanskrit and Prakrit poetry."[41]

3.5 THE SQUARE-ROOT OPERATION

The approximate value of the square root of the number 2 can be calculated using an arithmetical equation that is an alternative form of the Pythagorean theorem, as explained later in Section 3.8. This theorem is provided in several Śulbasūtras.[42] If we apply the expression to a square of side 1 in any unit system, we can find the value of $\sqrt{2}$. According to this theorem, the diagonal of a right-angle triangle with two equal sides (a) is

$$L = a + \frac{a}{3} + \frac{a}{3 \times 4} - \frac{a}{3 \times 4 \times 34}$$

For $a = 1$, this gives us the approximate value of $\sqrt{2}$.

$$L = \sqrt{2} = 1 + \frac{1}{3} + \frac{1}{3 \times 4} - \frac{1}{3 \times 4 \times 34} = \frac{577}{408}$$

$$\sqrt{2} = 1.41$$

This is the accepted value. If one tries to get higher accuracy in the results, the value is correct to the fifth decimal place, 1.41421. After the fifth decimal place, the value given in this formula is slightly higher (1.4142156) than the actual value (1.4142135).

Professor John F. Price of the University of New South Wales, Australia, provides the rationale of this formula in *Baudhayana Śulbasūtra*.[43] According to him, we know the value of $\sqrt{2}$ is between 1 and 2. If we equate $\sqrt{2}$ to 1 or 2 and square both sides we get 2 on one side while 1 or 4 on the other side. Using similar considerations, we know that $\sqrt{2}$ will be less than $(1 + \frac{1}{2})$ and more than $(1 + \frac{1}{3})$, Therefore, our first initial approximation is

$$\sqrt{2} \approx 1 + \frac{1}{3}$$

If we improve this value by adding a small term (x) to our value of $\sqrt{2}$, square both sides, assume x^2 to be quite small and neglect it, we go through the following sequence:

[41] Singh, 1985.
[42] *Baudhayāna-Śulbasūtra*, 2: 12; *Āpastambā-Śulbasūtra*, 1: 6; and *Kātyāyana-Śulbasūtra*, 2: 9.
[43] Price, in Gorini, 2000, p. 46–55.

$$\sqrt{2} = 1 + \frac{1}{3} + x = \frac{4}{3} + x$$

$$2 = \left(\frac{4}{3}\right)^2 + \frac{8}{3}x + x^2$$

$$2 = \left(\frac{4}{3}\right)^2 + \frac{8}{3}x$$

$$x = \frac{1}{3 \times 4}$$

In our next approximation,

$$\sqrt{2} \approx 1 + \frac{1}{3} + \frac{1}{3 \times 4} + y$$

This reduces to

$$\sqrt{2} = \frac{4}{3} + \frac{1}{3 \times 4} + y$$

Again, square and ignore y^2 term. The next step yields

$$2 = \frac{16}{9} + \frac{1}{9 \times 16} + \frac{2 \times 4}{9 \times 4} + y\left(\frac{2 \times 4}{3} + \frac{2}{3 \times 4}\right)$$

If we simplify this equation and calculate y, we get

$$y = -\frac{1}{3 \times 4 \times 34}$$

If we continue further, the next approximation will be

$$L = 1 + \frac{1}{3} + \frac{1}{3 \times 4} - \frac{1}{3 \times 4 \times 34} - \frac{1}{3 \times 4 \times 34 \times 2 \times 577}$$

This will give us $\sqrt{2}$ = 1.414213562374, a value accurate up to the thirteenth place. This approximation was done by David W. Henderson of Cornell University[44] and Price[45]. It is interesting that the *Śulbasūtra* does mention that the value is only approximate and a little bit higher than the actual value (*saviśeṣa*).[46] This process is called the method of successive approximations in most mathematics books, a modern technique.

[44]Read, Square Roots in the Śulba Sūtras by Henderson, in Gorini, 2000, p. 39–45.
[45]Price, John F., Applied Geometry of the *Śulba Sūtras*, in the book by Gorini, 2000, p. 46–55.
[46]Read, Square Roots in the Śulba Sūtras by Henderson, in Gorini, 2000, p. 39–45.

This process continues with more terms added. It always remains only an approximate value. Depending on the type of accuracy we need in everyday situation, the ancient Hindus felt appropriate to stop after reading to the fifth place of the decimal. Henderson also compares the popular "divide-and-average" method, also called the Newton's method, with the Baudhayāna's method and concludes that Baudhayāna's method "uses significantly fewer computations" than the Newton's method to reach the same order of accuracy.

3.6 ALGEBRA

Al-Khwārizmī, Muhammad ibn Mūsā is one of the earliest known astronomers and mathematicians from the glorious period of Baghdad/Islam. He was native to the Khwarizm Region in Persia, as the name suggests. He later moved to Baghdad and served in the court of al-Mamūn (813–833 CE). His best known works are: *Kitāb al-jabr wa'l Muqābala* (*The Book of Manipulation and Restoration*), *Kitāb al Hisāb al-Hindī* (*Book of Indian Mathematics*), and *Zīj al-Sindhind* (*Astronomy Table from India*). The last two books are obviously the works of the Hindus as the titles suggest. Even the third book was based on the work of the Hindus. The impact of al-Khwārizmī was so great in Europe that the title of his one book became synonymous with the theory of equations. The word "algebra" stemmed from the word *al-jabr* which appeared in the title of al-Khwārizmī's book. *Al-jabr* literally means bone-setting, indication manipulation of equations. Al-Khwārizmī played a crucial role as a disseminator of science. Latinization of his name took many forms: from *algorizmus* to *algoritmus* to *algorithmus*. The mathematical term, algorithm, has possibly stemmed from his Latinized name.[47]

John Wallis (1616–1703), an English mathematician, taught at Oxford University, wrote a book, *A Treatise of Algebra both Historical and Practical* in 1685. He is credited with introducing the symbol for infinity, ∞. In this book, he writes: "However, it is not unlikely that the Arabs, who received from the Indians the numerals figures (which the Greeks knew not), did from them also receive the use of them, and many profound speculations concerning them, which neither Latins or Greeks did know, till that now of late we have learned them from thence. From the Indians also they might learn their algebra, rather than from Diophantus, (who only of the Greeks wrote of it, and he but late, and in a method very different from theirs) . . . And the name they [Arabs] gave it (*Al-gjabr W'al-mokabala*) seems to have no affinity with any Greek name . . . "[48]

By the fifth century CE, the following are some of the rules that were known to the ancient Hindus. All these rules are taken from *Āryabhaṭīya*. Āryabhaṭa I did not provide the derivation of the rules in most cases and called the knowledge ancient.

[47]For more information, read Crossley and Henry, 1990; Hughes, 1989; King, 1983; Sesiano, in the book by Selin, 1997.
[48]Wallis, 1685, page 4, Chapter 2.

3.6.1 SUM OF A SERIES

"The desired number of terms minus one, halved, then increase by the number of the preceding terms (if any), multiply by the common difference between the terms, and then increase by the first term of the (whole) series. The result is the arithmetic mean (of the given number of terms). This multiplied by the given number of terms is the sum of the given terms. Alternatively, multiply the sum of the first and last terms by half the number of terms."[49]

This rule can be mathematically written for a series with initial term a and common difference between terms as d. mathematically, it can be written as

$$a + (a + d) + ...[a + (n - 1)d]$$

The sum of the series for n terms is:

$$S_n = n\left[\left(\frac{n-1}{2}\right)d + a\right]$$

$$= \frac{n}{2}[a + (a + (n - 1)d)]$$

$$= \frac{n}{2}[\text{first term} + \text{last term}]$$

These results are correct.

3.6.2 SUM OF A SERIES WITH Σn^2 AND Σn^3

"The continued product of the three quantities viz, the number of terms, number of terms plus one, and twice the number of terms increased by one when divided by 6 gives the sum of the series of squares of natural numbers. The square of the sum of the series of natural numbers gives the sum of the series of cubes of natural numbers."[50]

Mathematically, the above verse translates as follows:

For the series $1^2 + 2^2 + 3^2 + ... + n^2$, the sum equals to $\frac{n(n+1)(2n+1)}{6}$. This provides the correct value for all values of n.

Similarly, for the series $1^3 + 2^3 + 3^3 + ... + n^3$, according to Āryabhaṭa I, the sum equals to $(1 + 2 + 3 + ... + n)^2$. This reduces to $\left[\frac{n(n+1)}{2}\right]^2$ which provides correct result.

3.6.3 SOLUTION TO A QUADRATIC EQUATION

Āryabhaṭa I also gave the solution of a complex money-transaction problem: "Multiply the sum of the interest on the principal and the interest on this interest by the time and by the principal. Add to this result the square of the half of the principal. Take the square-root of this. Subtract

[49] *Āryabhaṭīya, Gaṇitpada*, 19.
[50] *Āryabhaṭīya, Gaṇitpada*, 22.

half the principal and divide the remainder by the time. The result will be the interest on the principal."[51]

The problem can be written as follows: "A certain amount of money (P for principal money) was given on loan for one month with unknown interest x. The unknown interest x that was accrued in one month was again loaned for time T months. On the completion of this period, the original interest x and the interest on this interest, all together, became I. Find out the rate of interest x on the amount P. The answer provided by Āryabhaṭa I is as follows:

$$x = \frac{\sqrt{IPT + (P/2)^2} - P/2}{T}$$

The solution provided by Āryabhaṭa I is correct.

3.7 GEOMETRY

The ancient Hindus used an elaborate knowledge of geometry for the construction of altars for religious purposes, for arranging various battalions of soldiers in wars, and for planning the design of their cities.[52] In scientific warfare, a smaller army can prevail with proper geometrical constructions in achieving their tactical goals. For example, Pāṇḍava in *Mahābhārata* used their much smaller army and these tactical geometrical constructions against a much larger army of Kaurava. *Cakravyūha* was an important geometrical construction of the placements of warriors to trap enemy warriors during the period of *Mahābhārata*. Abhimanyu, son of Arjuna, as mentioned in *Mahābhārata*, knew how to break and enter into the geometrical fortification of *cakravyūha*. However, he did not know the ways to come out of it. He was trapped and lost his life. In *Mahābhārata* war, multiple geometrical constructions, defined by their similarity to mostly animal or object shapes, were used in organizing soldiers to fight.

The word *śulba* means a chord, a rope, or a string and *Śulbasūtra* signifies geometry using strings. The *Śulbasūtra* are not the books of geometry or mathematics; these books deal mostly with rituals. The knowledge of geometry and mathematics is used to perform the rituals. One can consider them as the "first applied geometry text in the world" to "combine geometry and numerical techniques."[53]

3.7.1 TRANSFORMING A SQUARE INTO A CIRCLE

"If it is desired to transform a square into a circle, (a cord of length) half the diagonal (of the square) is stretched from the center to the east (a part of it lying outside the eastern side of the square); with one-third (of the part lying outside) added to the remainder (of the half diagonal), the (required) circle is drawn."[54]

[51] *Āryabhaṭīya*, *Gaṇitpada*, 25.
[52] Datta, 1932; Kulkarni, 1983; Sarasvati Amma, 1979; Sen and Bag, 1983; Staal, 1999; Thibaut, 1875.
[53] Henderson, in Gorini, 2000
[54] *Baudhayāna-Śulbasūtra*, 2: 9.

Let PQRS be the given square and O is the center of the square (see Figure 3.1). The half diagonal OP is drawn and an arc PAQ is formed, implying OP = OA. A circle of radius OB (r) is drawn where OB = OC + CB and CB = $\frac{1}{3}$CA.

This implies that OB = OC + $\frac{1}{3}$CA.

This equals OB = OC + $\frac{1}{3}$ (OA - OC).

Let $2a$ be the side of square PQRS,

$$r = a + \frac{1}{3}(\sqrt{2}a - a)$$

$$r = a[1 + \frac{1}{3}(\sqrt{2} - 1)]$$

$$r = \frac{a}{3}(2 + \sqrt{2})$$

The area of the circle is $\pi r^2 = 3.14[\frac{a}{3}(2 + \sqrt{2})]^2 = 4.06a^2$. This is in rough agreement with the area of the given square ($4a^2$).

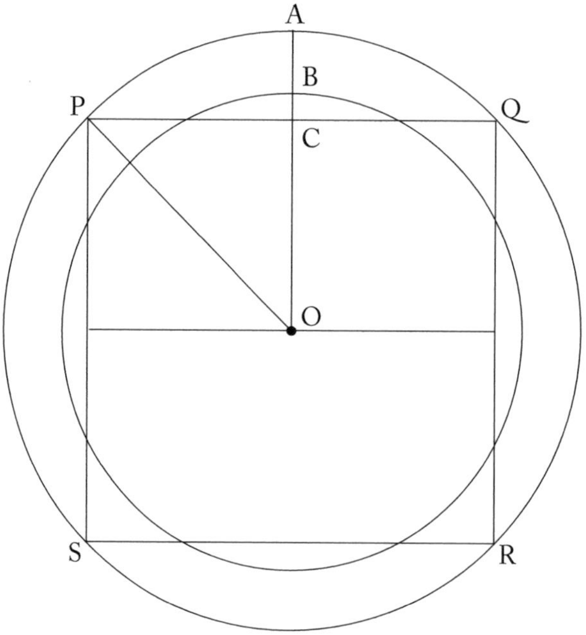

Figure 3.1: Transforming square into a circle of the same area.

Hindus' mathematical books did not provide axioms and postulates; the mathematical rules are based on experimental verification. As a result, some rules are only approximately correct. Seidenberg tried to assign a possible period for the general use of geometry among the Hindus but realized, "no matter how far we go back in 'history', we find geometrical rituals."[55]

3.7.2 HEIGHT OF A TALL OBJECT

"(When two gnomons of equal height are in the same direction from the lamp-post), multiply the distance between the tips of the shadows (of the two gnomons)[CD] by the (larger [GD] or shorter [EC]) shadow and divide by the larger shadow diminished by the shorter one [GD - EC]: the result is the upright (i.e., the distance of the tip of the larger [BD] or shorter shadow [BC] from the foot of the lamp-post). The upright multiplied by the height of the gnomon and divided by the (larger or shorter) shadow gives the base (*i.e.* height of the lamp-post)."[56]

Mathematically,

$$BD = \frac{CD \times GD}{GD - EC}$$

and

$$AB = \frac{h \times BC}{EC} = \frac{h \times CD}{GD - EC}$$

The proof of this theorem is as follows (see Figure 3.2). Let AB be the unknown height that we want to find out. We place a gnomon of height h at point E. This gives us the shadow EC. Now we place the same gnomon or another one of the same size at point G. This gives us shadow GD.

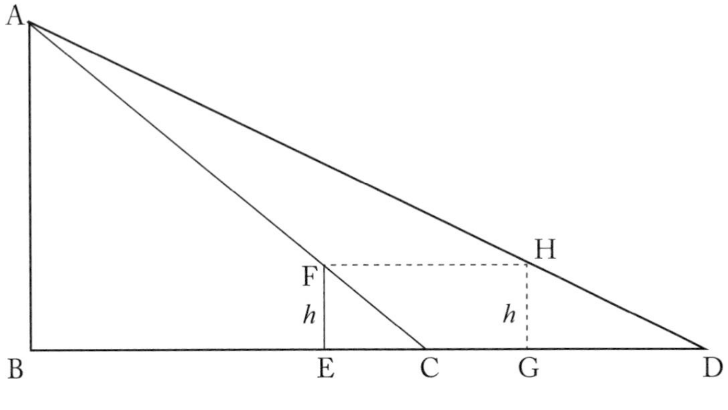

Figure 3.2: Determining height of a tall object using a small stick.

Using similar triangles,

$$\frac{AB}{h} = \frac{BC}{EC} = \frac{BD}{GD}$$

[55] Seidenberg, 1962.

[56] *Āryabhaṭīya*, *Gaṇitapada*,16. For more information, read Mishra and Singh, 1996; and Shukla and Sarma, 1976, p. 58.

$$\text{AB} = \frac{\text{BC} \times h}{\text{EC}}$$

As given before,

$$\frac{\text{BC}}{\text{EC}} = \frac{\text{BD}}{\text{GD}}$$

Since, BD = BC + CD,

$$\frac{\text{BC} + \text{CD}}{\text{GD}} = \frac{\text{BC}}{\text{EC}}$$

$$\frac{\text{BC} + \text{CD}}{\text{BC}} = \frac{\text{GD}}{\text{EC}}$$

$$1 + \frac{\text{CD}}{\text{BC}} = \frac{\text{GD}}{\text{EC}}$$

$$\frac{\text{CD}}{\text{BC}} = \frac{\text{GD}}{\text{EC}} - 1$$

$$\frac{\text{CD}}{\text{BC}} = \frac{\text{GD} - \text{EC}}{\text{EC}}$$

$$\text{BC} = \frac{\text{CD} \times \text{EC}}{\text{GD} - \text{EC}}$$

Therefore,

$$\text{AB} = \frac{h \times \text{BC}}{\text{EC}} = \frac{h \times \text{CD}}{\text{GD} - \text{EC}}$$

Thus, by measuring the lengths of shadows at two different locations for a stick, one could measure the height of the unknown object. Or, by aligning the stick in such a way that one could see the top of the object in line with the top of stick while looking from the ground level from two different locations, one could measure the height of the unknown object.

3.7.3 THE VALUE OF π

The Greek letter π (pronounced as *pi*) indicates the ratio of the circumference of a circle to its diameter that is a constant for any circle. The ancient Hindu books generally provide two values of π: one for rough calculations and second for precise measurements. The knowledge of π is useful in the construction of altars, wheels of a cart, the metallic rims of a wheel, and in geometry.

The *Mānava-Śulbasūtra* provided the value of π to be 3.2: "The fifth part of the diameter added to the three times the diameter gives the circumference (of a circle). Not a hair of length is left over."[57] This provides the value of circumference, C, from diameter, D, as,

$$C = \frac{D}{5} + 3D = \frac{16}{5}D = 3.2D$$

This gives a value of π which is close to the actual value of 3.14. Since the purpose of these books was to prepare altars, these calculations were good enough for that purpose.

Āryabhaṭa I gave the value of π that is correct to the fourth decimal place: "Add four to hundred, multiply by eight, and add sixty two thousand. The result is approximately the circumference of a circle [C] of which the diameter [D] is twenty thousand."[58]

Mathematically, we know that $C = \pi D$. Therefore, the value of π, based on Āryabhaṭa I's method, is equal to:

$$\pi = \frac{[(4 + 100) \times 8] + 62,000}{20,000}$$

$$= \frac{62,832}{20,000} = 3.1416$$

This is equal to the presently accepted value of 3.1416, for up to 4 decimal places. Āryabhaṭa I called this value to be approximate. This makes sense since the value of π can only be determined approximately since the ratio of circumference to diameter is not evenly divisive; it can have an innumerable number of significant figures. It is an endeavor for many mathematicians to calculate a more precise value of π. The value of π to a large number of significant figures is commonly used to check the speed, efficiency, and the accuracy of computers.

3.8 THE PYTHAGOREAN THEOREM

The so-called Pythagorean theorem connects the three sides of a right-angle-triangle with the relation,

$$a^2 + b^2 = c^2$$

where a, b, and c are the base, perpendicular, and hypotenuse, respectively of a right-angle triangle (see Figure 3.3). Pythagoras, a Greek philosopher is said to be the originator of the theorem sometime during sixth century BCE

Let us provide a few statements from the *Śulbasūtra* that indicate the understanding of the Pythagorean theorem among the Hindus:

[57] *Mānava-Śulbasūtra*, 11: 13.

[58] *Āryabhaṭīya*, *Gaṇitapada*, 10. Also, Read Hayashi *et al.*, 1989 and Kulkarni, 1978a. The work of Hayasi *et al.* has a good review of the later developments on the issue in India. For example, Madhav (14th century) calculated the value of π that was correct to eleven places.

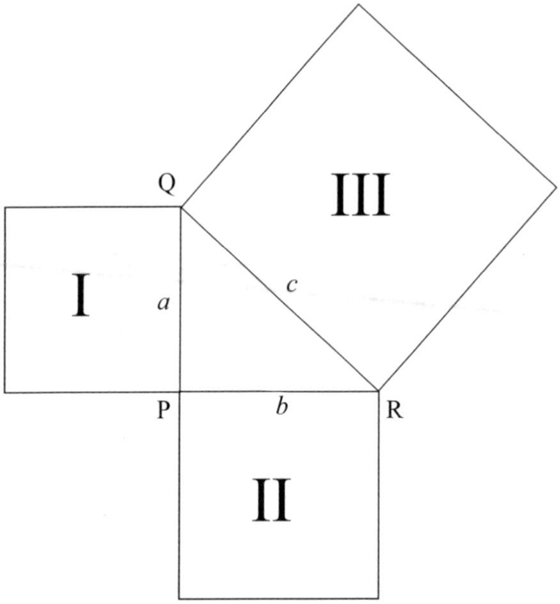

Figure 3.3: A right-angled triangle, PQR.

"The diagonal of a square produces double the area (of the square). It is $\sqrt{2}$ (*dvikaraṇī*) of the side of the square (of which it is the diagonal)."[59] The word *dvikaraṇī* literally means "that which produces 2," implying the twice area of the square constructed using *dvikaraṇī* as one side in comparison to the original square.

"The diagonal (of a right angle triangle) of which the breadth is *pada* and the length 3 *padas* is $\sqrt{10}$ *padas*."[60]

"The diagonal (of a right triangle) of which the breadth is 2 *pada* and the length is 6 *pada* is $\sqrt{40}$ *pada*."[61]

The *Baudhayāna-Śulbasūtra* provides another rule for a right-angle triangle: "The areas (of the squares) produced separately by the length and the breadth of a rectangle together equal the area (of the square) produced by the diagonal."[62]

Similar rules are given elsewhere in the ancient Hindu literature: "The areas (of the squares) produced separately by the length and the breadth of a rectangle together equal the area (of the square) produced by the diagonal. By the understanding of these (methods) the construction of the figures as stated (is to be accomplished)."[63]

[59] *Āpastambā-Śulbasūtra*, 1: 5.
[60] *Kātyāyana-Śulbasūtra*, 2: 4.
[61] *Kātyāyana-Śulbasūtra*, 2: 5.
[62] *Baudhayāna-Śulbasūtra*, 1: 48.
[63] *Āpastambā-Śulbasūtra*, 1: 4.

There is a curious similarity between Baudhayāna and Pythagoras on the so-called Pythagorean theorem. It is important to understand that Baudhayāna's work was compiled at least around 1700 BCE, while the Pythagorean theorem was compiled over a millennia later, i.e., around sixth-century BCE.[64] There are indications that Pythagoras perhaps came in contact with India or Indian wisdom.

The Pythagorean theorem is mostly given in terms of geometry in most cultures. However, among the Hindus, they used a mathematical expression for the Pythagorean theorem which is unique to them. Several *Śulbasūtras* provided a arithmetical method to find the diagonal of a right-angled triangle with two equal sides: "The measure is to be increased by its third and this (third) again by its own fourth less the thirty-fourth part (of the fourth); this is (the value of) the diagonal of a square (whose side is the measure).[65]

Mathematically, for a right-angled triangle with two equal sides of length a, the hypotenuse is equal to

$$L = a + \frac{a}{3} + \frac{a}{3 \times 4} - \frac{a}{3 \times 4 \times 34}$$

No other culture has provided a similar mathematical form to calculate the hypotenuse.

3.9 TRIGONOMETRY: FROM *JYĀ* TO SINE

Trigonometry is a branch of science which deals with specific functions of angles and their application to calculations in geometry. The sine function, as defined in trigonometry, is essential to the study of geometry. For a right-angle-triangle, if θ is the acute angle of a right triangle, the mathematical symbol sin, pronounced as *sine*, of θ is the ratio of the side opposite (b) and the hypotenuse (c). Mathematically, $\sin\theta = \frac{b}{c}$.

Why do we call this function *sine*? Who chose this word for the scientific community for what reason? What is the meaning of this word? These are simple questions that intrigue curious minds when they first learn about this trigonometric function. The answer is as follows: The sine function was called *jyā* by Āryabhaṭa I. Al-Khwārizmī, when borrowed this concept, chose an Arabic that is similar in pronunciation. He used *geib* or *jaib* for this term. This word has a specific meaning: fold or pocket. The Latin word for pocket or fold is sinus, and, thus, the term sine for this trigonometric function evolved.

"The most significant contribution of India to medieval mathematics is in trigonometry. For a circle of unit radius the length of an arc is a measure of the angle it subtends at the center [center] of the circle. The Greeks, to facilitate calculations in geometry, tabulated values of the chord of arcs. This method was replaced by Hindu mathematicians with half chord of an arc, known as sine of the angle . . . No influence on the West was exerted by the development in India . . . Thus, methods had been known in India were not rediscovered until 1624 by the

[64] Seidenberg, 1962. It is likely that the date assigned by Seidenberg will be revised with new scholarship.

[65] *Baudhayāna-Śulbasūtra*, 2: 12; *Āpastambā-Śulbasūtra*, 1: 6; and *Kātyāyana-Śulbasūtra*, 2: 9.

French mathematician Claude-Gaspar Bachet, sieur de Méziriac,"[66] suggests the *Encyclopedia Britannica* (see Figure 3.4).

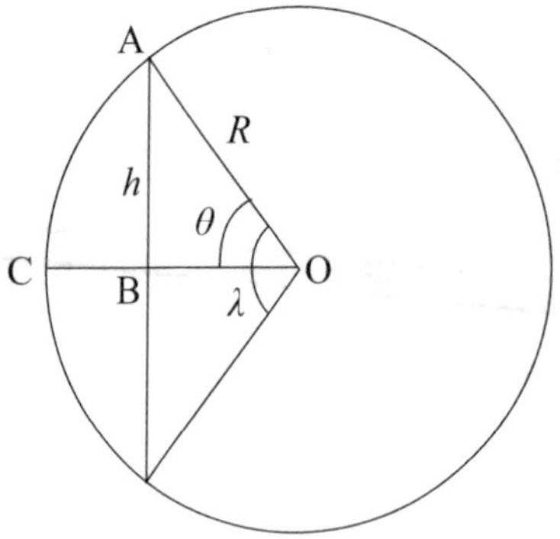

Figure 3.4: Āryabhaṭa I's method to define an arc.

Āryabhaṭa I used the half chord on an arc, defined the *sine* function and gave a table of *sines* of the angles. In his book, the circumference of a circle was divided into $360 \times 60 = 21,600$ equal segments. He also divided the quadrant of a circle into 24 equal parts. The smallest is thus $225'$ or $3°45'$. Āryabhaṭa I defined *jyā* (or $\sin\theta$) for a circle of radius, R, of 3438 units. What is the mystery behind this strange number 3438? Well, we know that there are 21600 minutes of arc in a circle of 360 degrees. If we divide this number by 2π, we get 3437.74 = 3438. A consequence of this is that if angle θ is measured in minutes of arc and is small, then $3438 \sin\theta$ is approximately equal to θ. For example, for a 5-degree angle, the angle comes out to be 300 in minutes. $R\sin\theta$ = 3438 sin (300) = 299.6 = 300, implying that R = 3438 is similar to using radians instead of degrees in modern mathematics, where we have that $\sin x$ is approximately equal to x.[67] In Table 3.2, a comparison of Āryabhaṭa I's values with the modern values is provided.

3.10 DIFFUSION OF HINDU MATHEMATICS TO OTHER CULTURES

3.10.1 THE MIDDLE EAST

Owing their gratitude to the Hindus, numerals were always called *arqam hindiya* in the Middle East, meaning the Hindu numerals.[68] These numerals were known as Hindu numerals throughout the medieval period in the Arab world and are known so even today. Al-Jāḥiz, (ca. 776–868 CE), al-Khwārizmī (ca.800–847 CE), al-Uqlīdisī (ca. 920–980 CE) and Ibn Labbān (ca. 971–1029 CE), several famous natural philosophers from the Middle East or nearby regions, have testified to the Hindu origin of the so-called Arabic numerals.

[66] *Encyclopedia Britannica*, vol. 23, p. 605, 1989.
[67] For more details, read Achar, 2002; Clark, 1930; Hayasi, 1997; and Shukla and Sarma, 1976.
[68] Sarton, 1950.

Table 3.2: Āryabhaṭa I's and Modern *sine* values

Angle	Āryabhata I's Value	Modern *sine* value × 3438
3°45'	225	225
7°20'	449	449
11°15'	671	671
15°0'	890	890
18°45'	1105	1105
22°30'	1315	1315
26°15'	1520	1521
30°0'	1719	1719
33°45'	1910	1910
37°30'	2093	2092
41°15'	2267	2266
45°0'	2431	2431
48°45'	2585	2584
52°30'	2728	2728
56°15'	2859	2859
60°0'	2978	2977
63°45'	3084	3083
67°30'	3177	3176
71°15'	3256	3256
75°0'	3321	3321
78°45'	3372	3372
82°30'	3409	3409
86°15'	3431	3431
90°0'	3438	3438

Interestingly, in conformity with Arabic tradition, these numerals were called Hindu all through the medieval and early Renaissance periods in Europe by their top scholars. Adelard of Bath (1116–1142 CE) and Roger Bacon (1214–1292 CE) in England, Leonardo Fibonacci (1170–1250 CE) in Italy, Ṣā'id Al-Andalusī (1029–1070 CE) and Ibn Ezra (11th century CE) in Spain, and Voltaire (1694–1778 CE) in France called them Hindu numerals. They were labeled Arabic only after the sixteenth century by some.[69] This was the beginning of the colonial period. Why the Hindu numerals got changed to Arabic numerals is a mystery in the history of mathematics. Did it happen due to a coincidence or was it a result of complex political ambitions of the West? This is a curious mystery that is not yet resolved. Today, most textbooks call these numerals the Hindu-Arabic numerals.

Al-Jāḥiz, (776–868 CE), a well-known Arab theologian and natural philosopher, attributed the origin of numerals to the Hindus.[70] Using Hindu numerals, Kūshyār Ibn Labbān (ca. 971–1029 CE) wrote a book on Hindu arithmetic, *Principles of Hindu Reckoning*,[71] which became quite popular during the eleventh-century Islamic world. Kūshyār ibn Labbān was primarily an astronomer from Jilan, a village in Persia on the south of Caspian Sea, and wrote his manuscripts in Arabic. He started his book with the following sentence: "In the name of Allāh, the Merciful and Compassionate. This book on the principle of Hindu arithmetic is an arrangement of Abu al-Hasan Kushyār bin [Ibn] Labbān al-Jīlī, may God have mercy on him."[72]

In the time of communal tensions around the world, Labbān's book sets an example for all—a Muslim who wrote a book on the Hindu arithmetic with the mercy of Allāh. It is completely in line with the doctrines of Prophet Mohammad who suggested that his followers travel around the world for knowledge: "You should insist on acquiring knowledge even if you have to travel to China."

Ibn Sīnā (980–1037 CE), a noted Persian scholar of the eleventh century mentions that the Hindu method of calculation was common in Persia during his period. Even merchants with small businesses used the Hindu mathematical methods for calculations. Ibn Sīnā himself learned such methods from a vegetable merchant, Mahmūd al-Massāhī.[73] Ibn Sīnā was a prolific writer with books on medicine, metaphysics, natural philosophy, and mathematics. He explained the Greek works of Aristotle, Plato, and Plotinus within the framework of Islam.

Since the Greek numerals were already in use in Arabia when the Hindu numerals were introduced, both numeral systems competed with each other. Lobbies were formed where the superiority of one system over the other was debated/discussed. Severus Sebokht (died, 662 CE), a Syrian natural philosopher mentioned of such a rivalry between the Greek and Hindu numerals. [74]

"I will omit all discussion of the science of the Hindus, a people not the same as Syrians, their subtle discoveries in the science of astronomy, discoveries which are more ingenious than those of the Greeks and the Babylonians; their valuable method of calculation; their computing

[69] Clark, 1929, p. 217.
[70] Pellat, 1969, p. 197; Levy and Petruck, 1965, p. 6.
[71] Levy and Petruck, 1965.
[72] Levy and Petruck, 1965.
[73] Gohlman, 1974, p. 21 and 121.
[74] Wright, 1966, p. 137. Sebokht was born in Nisibis in Persia and taught Greek philosophy in Syria. Later, he became the bishop of the convent of Ken-neshre (Qenneshre), a center of Greek learning in Upper-Euphrates in West-Syria. He had a good grasp of various sciences and wrote on astronomy, geography, and mathematics. His manuscripts are preserved in museums in London and Paris. Sebokht believed in the universal nature of science and was of the opinion that many cultures, not just the dominant Greeks, contributed to the Arab scientific pool.

that surpasses description. I wish only to say that this computation is done by means of nine signs. If those who believe, because they speak Greek, that they have reached the limits of science, should know these things, they would be convinced that there are also others [Hindu] who know something."[75] This quotation is a proof that the Hindu numerals were in practice in Arabia by the seventh-century. Also, by the seventh century, people started to discard the Greek numerals in favor of the Hindu numerals in Arabia.

Al-Uqlidīsī (920–980 CE, also written as Uklidisi, meaning 'the Euclid-man' for his role as copyist of Euclid's work) in his book, *Kitāb al-Fusūl fī al-hisāb al-Hindī* defined the Hindu numeral system and explained Hindu arithmetic: "I have looked into the works of the past arithmeticians versed in the arithmetic of the Indians [Hindus, in the original] . . . We can thus dispense with other works of arithmetic. The reader who has read other works will realize this fact and thus adhere to this work and prefer it to others, new or old, for it contains more than any other books of this kind."[76] The title of the above book indicates that the book contained the mathematics of the Hindus. According to al-Uqlidīsī, "even the blind and the weak sighted will find in our explanations and summaries something to benefit from without toil or cost, by the will and help of God."[77] He mentioned the use of the dust-board abacus for the Hindu arithmetic [*hisāb al-Ghubār*, *hisāb al-takht* (board), *al-Hisāb al-Hindī* or *Hisāb al-turāb* (dust)].[78] This kind of board became so popular for mathematical calculations, arithmetic was called *Hisāb al-Ghubār*, meaning mathematics of the dust-board. This term was common in the medieval Spain.[79] On the practicality of a dust board in mathematical calculations, al-Uqlidīsī writes, "If some persons dislike it because it needs the *takht* (board), we say that this is a science and technique that needs a tool."[80]

Al-Khwārizmī wrote a book in the Arabic language on the use of Hindu numerals, *Kitāb al-hisāb al-Hindī*. Although the original book in the Arabic language has been lost, its Latin translation by Robert of Chester has survived.[81] A thirteenth century manuscript from Spain has been found recently which is more complete than the first Latin translation. This manuscript along with its German translation was published by Menso Folkerts in 1997.[82] Al-Khwārizmī's book is a compilation of the mathematical procedures used by the Hindus, as acknowledged by himself.[83] Adelard of Bath (1080–1152 CE), an English philosopher, mathematician, and

[75] Datta and Singh, pt. 1, p. 96; Clark, 1929, p. 220.

[76] Saidan, 1978, p. 35.

[77] Saidan, 1978, p. 189. This book was composed in Damascus in 952–953 CE and a 1186 CE manuscript has survived. Read Burnett, 2006.

[78] Saidan, 1978, p. 12.

[79] Read, Salem and Kumar, 1991.

[80] Saidan, 1978, p. 35; Saidan has also compiled a list of books on Hindu-Arabic arithmetic that were in the Arab world. See, Saidan, 1965.

[81] Hughes, 1989.

[82] Folkerts, Menso, *Die älteste Lateinische Schrift über das Indische Rechnen nach al-Ḫwārizmī*, Übersetzung und Kommentar von M. Fokerts under Mitarbeit von Paul Kunitzsch, Bayerische Akademie der Wissenschaften, Philosophisch-historische Klasse, Abhandlungen, Neue Folge, Heft 113, 1997. Taken from Folkerts, 2001.

[83] Folkerts, 2001. Al-Khwārizmī was perhaps the most influential Islamic philosopher in Europe before the European Renaissance. The mathematical terms sine, algorithm, and zero are attributed to him.

scientist, translated al-Khwārizmī's astronomical tables, that had been previously revised by al-Majrīṭī (d. 1007 CE), from Arabic to Latin. These tables included tables of *sine*, and, thus, the *sine* function was introduced to the Latin world.[84] Further, this Latin translation was translated into English by O. Neugebauer.[85] It is this English translation that is our major source of al-Khwārizmī's knowledge in astronomy and trigonometry.

Trigonometry was introduced to the Arab world by India.[86] Al-Battānī, Abu 'Abd Allāh Muhammad ibn Jābir ibn Sinān al-Raqqī al-Harrānī al-Sābi' (858–929 CE) used the half-cord (that leads to *sine* function) instead of the chord in his book *Kitāb al-Zīj*(Book of Tables) following the examples of his predecessors who used the Hindu method rather than the Greek method.[87] Al-Battānī's *Zīj* was translated into Latin by Robert of Ketton and Plato of Tivoli during the twelfth century. This table was also translated into Spanish under the patronage of Alfonso X.[88]

3.10.2 CHINA

Yoshio Mikami, in his book, *The Development of Mathematics in China and Japan*, wrote of the Indian influence on Chinese mathematics: "Things Indian exercised supremacy in art and literature, in philosophy, in the mode of life and the thoughts of the inhabitants, in everything. It is even said, astronomy and calendrical arts had also felt their influence. How then could arithmetic remain unaffected? No doubt the Chinese studied the arithmetical works of the Hindoos [Hindus]."[89] On the popularity of Indian literature in China, Mikami wrote: "we read in history, the Indian works were read in translation ten times more than the native [Chinese] classics, a fact that vividly tells how the Indian influence had swept over the country [China]."[90]

The biggest export of India to China is definitely the religion founded on the teaching of Lord Buddha, Buddhism. When Buddhism was introduced to China, this country was already an old civilization with a powerful tradition and history. Philosophers such as Confucius guided Chinese society with their philosophies. Confucianism mostly deals with ethical rules that are applicable to this world; it does not deal with the spiritual life. Buddhism filled that niche to allow people to think about the "ultimate questions" of life, liberation, etc. There was a contrast in the intentions of these visitors who came from China in comparison to the Europeans. Most Chinese travelers visited India for their spiritual pursuits while Marco Polo and Vasco de Gamma visited to collect wealth. For example, Xuan Zang (b. 603 CE, also known as Hiuen Tsang, Huan Chwang, Yuan Chwang, Hiouen Thsang, and Hsuan Tsang)[91] studied in Nālandā and carried books on twenty-two horses on his way back to China. He build a pagoda in Xian,

[84]Sarton, 1931, vol. 2, p. 167.
[85]Neugebauer, 1962.
[86]Needham and Ling, 1959, vol. III, p. 108.
[87]*Dictionary of Scientific Biographies*, vol. I, p. 508
[88]Julio Samsó in Selin, 1997
[89]Mikami, 1913, p. 57.
[90]Mikami, 1913, p. 57.
[91]Bernstein, 2001, p. 22.

China to house the books. Kumārjīva, an Indian scholar, went to Chang-an in 401 CE and served as "Grand Preceptor" of an enormous translation project involving thousands of monks and scholars that advanced the philosophy of the great Indian philosopher Nāgārjuna.[92] He was from the Andra Pradesh region and was the founder of the *Madhyāmika* (Middle Path) school of Buddhism. This school became popular in China under the name *Sanlun*. Nāgārjuna was based in Nālanda University.

Ch'u-t'an Hsi-ta from Tang's Court translated a Sanskrit text into Chinese and introduced decimal notation and the arithmetic rules during the early eighth century. His work was continued further by Yijing (or I-tsing, 643–713 CE), under the order of the emperor.[93] (See Chapter 2.)

The trigonometric term *jyā*, as defined by Āryabhaṭa I, was called *ming* in China.[94] In the section, *Tui yueh Chien Liang Ming* (On the prediction of the Moon's positions) of *Chiu-Chih-li*, an astronomical book that is based on Sanskrit works, *chia* is the term used in this book to define the trigonometric function sine which is an obviously a minor change in pronunciation of the term *jyā*.[95] The term *ming* may have been adopted from the title of this book later as the book became popular in China. In 718 CE, a new calender, called *Jiuzhi li* (Nine Upholder Calendar), was compiled in China.[96] This calendar is based on Varāhamihira's *Pañca Siddhānta* and contains a table of sines and the Hindu numeral zero in the form of a dot (*bindu*).[97]

3.10.3 EUROPE

Monk Vigilia of the monastery of Albelda in the Rioja, Asturias (an autonomous community in Spain), in 976 CE, made a copy of the *Etymologies* that was originally written by Isidore of Seville (560–636 CE). In this copy, the monk provided information about the Hindu numerals with the following remark: "We must know that the Indians have a most subtle talent and all other races yield to them in arithmetic and geometry and the other liberal arts. And this is clear in the nine figures with which they are able to designate each and every degree of each order (of numbers)."[98]

Ṣā'id al-Andalusī (1029–1070 CE), a Spanish scholar of the eleventh century, writes about the Hindu numerals and arithmetic: "That which has reached us from their [Hindus'] work on numbers is *Hisāb al-Ghubār* (dust board arithmetic) which was simplified by Abū Jā'far Muhmmad ibn Mūsā al-Khwārizmī. This method of calculating is the simplest, fastest, and easiest method to understand and use and has a remarkable structure. It is a testimony to the intelligence of the Hindus, the clarity of their creativity, and the power of their inventiveness."[99]

[92]Bernstein, 2001, p. 95
[93]Sarton, 1927, vol. 1, p. 513.
[94]Martzloff, 1997, p. 90 and 100.
[95]Sen, 1970
[96]Yoke, 1985, p. 162
[97]Yoke, 1985, p. 162.
[98]Burnett, 2006.
[99]Salem and Kumar, 1991, p. 14.

Ṣā'id al-Andalusī was not the only one in Spain to write about the dominance of India in the sciences. Rabbi Abrahm Ibn Ezra (1096–1167 CE), a poet, scholar, and author of numerous books on grammar, philosophy, medicine, astronomy, astrology and mathematics, wrote on the ways the Hindu numerals were introduced in Arabia. Ibn Ezra mainly lived in Spain and translated al-Bīrūnī's and al-Khwārizmī's work into Hebrew. Ibn Ezra mentions the role of the Jewish scholars in bringing a Hindu scholar, named Kanaka, who taught place-value notations to the Arabs.[100]

Leonardo Fibonacci (1170–1250 CE) is well known for his contributions in arithmetic and geometry. He is also known for the Fibonacci Sequence. He learned the Hindu mathematics from the Arabs and popularized Hindu numerals in the western world. "Of all the methods of calculation, he [Fibonacci] found the Hindu [method] to be unquestionably the best,"[101] concludes Cajori in his studies. Fibonacci attributed the numerals correctly to the Hindus. Fibonacci wrote a statement in this book which erases any confusion about his knowledge of the Hindu numerals: "The nine Indian figures are: 9 8 7 6 5 4 3 2 1. With these nine figures, and with the sign 0 . . . any number may be written, as is demonstrated below."[102] He further writes to explain the place value notation: "The first place in the writing of the numbers begins at the right. The second truly follows the first to the left. The third follows the second. . . . the nine figures that will be in the second place will represent as many tens . . . the figure that is in the third place denotes the number of hundreds. . . if the figure four is in the first place, and the unit in the second, thus 14, undoubtedly xiiii will be denoted . . if the figure of the unit is in the first place, and the figure four in the second, thus 41, xli will be denoted."[103]

Fibonacci was fascinated with Hindu mathematics. ". . . the art of the nine Indian figures, the introduction and knowledge of the art pleased me so much above all else, and I learned from them, whoever was learned in it, from nearby Egypt, Syria, Greece, Sicily and Provence, and their various methods, to which locations of business I traveled considerably afterward for much study, and I learned from the assembled disputations," writes Fibonacci.[104]

Leonardo used the mathematical knowledge of the Hindus for business transactions from one currency to another, investment of money, calculation of simple and compound interest, and in defining the rules of barter. Fibonacci found the prevalent Roman numerals inferior to the Hindu numerals. The purpose of writing his book *Liber Abaci* was to introduce the Romans to the very best available tools of mathematical calculations and their applications.

Fibonacci started the book with the following sentence: "You, my Master Michael Scott, most great philosopher, wrote to my Lord about the book on numbers which some time ago I composed and transcribed to you; whence complying with your criticism, your more subtle examining circumspection, to the honor of you and many others I with advantage corrected this

[100] Goldstein, 1996.
[101] Cajori, 1980, p. 121.
[102] Boncompagni, 1857–1862; Horadam, 1975.
[103] Sigler, 2002, p. 17–18.
[104] Sigler, 2002, p. 15–16.

work. In this rectification I added certain necessities, and I deleted certain superfluities. In it I present a full instruction on numbers close to the method of the Indians whose outstanding method I chose for this science."[105]

Roger Bacon (1214–1294 CE), a noted Franciscan natural philosopher from England, studied at Oxford University and the University of Paris. Several manuscripts about the Hindu mathematics were available in French at the University of Paris that were read by Bacon.[106] Bacon emphasized the role of mathematics for Europe in his *Opus Majus* and deemed it "necessary to the culture of the Latins"[107] Fibonacci's *Liber Abaci* was published thirteen years prior to the birth of Bacon. As an adult, this book became known to Bacon. Bacon suggested that people should study mathematics from "all the sages of antiquity" and chastised them for being "ignorant" of its "usefulness."[108] Bacon felt that mathematics is "the gate and key" to learn other sciences. In his view, a study of mathematics was essential for a study of philosophy that, in turn, was essential for the study of theology. Therefore, mathematics "has always been used by all the saints and sages more than all other sciences,"[109] and is essential for theology and to know about the creatures and other creations.[110] Thus, mathematics, in the mind of Roger Bacon, was a tool to know more about theology or to know about God's creations. This is similar to how Nārada viewed mathematics to achieve salvation (Chapter 2).

The triumph of the Hindu numeral system over the Roman numerals precipitated neither rapidly nor without conflict. It was not easy to discard the Roman numerals as they were deeply rooted in the system. Initially, the new Hindu system was mixed with the Roman by some where place value notation was practiced using Roman symbols. For example, 1482 was written as M.CCCC.8II, whereas 1089 was written as I.0.VIII.IX.[111] In the twelfth and early thirteenth century, it was common to write numbers below 360 in Arabic, Greek or Latin symbols and the numbers above 360 in Hindu numerals.[112] In 1299 CE, the City Council of Florence forbade the use of Hindu numerals in official accounting books. Similarly, as late as the fifteenth century, the mayor of Frankfurt issued an order to the master calculators to abstain from the use of the Hindu numerals.[113] Some even tried to create a parallel system with the first nine Roman numerals. For example, H. Ocreatus, a student of Adelard of Bath, tried to create this system which could have been appropriate scientifically. However, it did not catch on among the scholars.[114] As late

[105] Sigler, 2002. Michael Scot was Leonardo's mentor.

[106] Smith, a chapter, The Place of Roger Bacon in the History of Mathematics, in the book by Little, 1914, p. 156.

[107] Bacon, 1928, vol. 1, p. 27.

[108] Bacon, 1928, vol. 1, p. 27.

[109] Bacon, 1928, vol. 1, p. 116.

[110] Bacon, 1928, vol. 1, p. 195.

[111] Gupta, 1983; Menninger, 1970, p. 287. In the number 1482, an amalgamation of the Roman and Hindu system is practiced. The number 1 represents 1000 and M is the corresponding symbol. Similarly, C represents 100. The second number 1089 is written in place-value notations with Roman symbols for different numbers. As mentioned earlier, symbols can easily be changed without compromising quality.

[112] Burnett, 2006.

[113] Gazalé, 2000, p. 48.

[114] Burnett, 2006

as the eighteenth century, the National Audit Office (*Cour des Comptes*) in France was still using Roman numerals in their accounting.[115]

As mentioned earlier, the Arabic term *geib* or *jaib* is a metamorphosed form of the term *jyā* used by the Hindus. The Arabic word *jaib* has multiple meanings: pocket, fold, or bosom. Adelard of Bath used the term *elgeib* for the trigonometric function *sine*, following the word used by al-Khwārizmī who use *geib* or *jaib* for this term.[116] Rabbi Ben Ezra (born c. 1093), of Toledo, Spain, also used a similar term, *al-geib*, for this trigonometric function.[117] Eventually, Gherardo of Cremona (ca. 1114–1187 CE, sometimes also spelled as Gerard or Gerhard, born in Carmona, Spain) literally translated the Arabic term into Latin and used the term *sinus* to define the operation.[118] The term *sinus* means bosom, fold, or pocket in Latin. Pocket or bosom has nothing to do with the trigonometric function. However, this term has been in use for about a millennium.

In 1631, Richard Norwood (1590–1675), a British mathematician and surveyor, published a book on trigonometry where the word *sine* was used for this trigonometric function and the symbol "*s*" was used to depict the function in his mathematical equations.[119] In 1634, the French mathematician Pierre Hérigone (1580–1643) became the first person to use the symbol "sin" for *sine* and the practice has continued since then.[120] David Eugene Smith has provided an excellent review of sine function.[121]

3.10.4 SUPPORT OF POPE SYLVESTER II

Pope Sylvester II was instrumental in popularizing the Hindu numerals, in his life as Gerbert d'Aurillac (945–1003 CE), his name prior to becoming pontiff. He was born in a poor family in France and was first educated at Aurillac, France. He later moved to Catalonia, Spain. He visited the cities of Barcelona, Seville, and Cordova in Spain which were leading centers of learning during the period. Gerbert became the headmaster of the cathedral school at Reims, later became the archbishop of Reims and then of Ravenna in Italy. His education in France and Spain in Latin grammar, science, and mathematics and his political astuteness, allowed him to become the hundred forty-sixth pope. With the patronage of Otto III of Saxony, on April 9, 999, he was elected as pontiff and chose Pope Sylvester II as his new name. It was an important period for Christianity as it was about to complete one thousand years since the birth of Christ. Pope Sylvester II is known for his efforts to translate Greek and Arabic texts on natural philosophy into Latin. He raised the profile of natural philosophy and reinforced intellectual aspects of theology in the mainstream activities of the Church.

[115] Sarton, 1950.
[116] Neugebauer, 1962, p. 44, 45.
[117] Goldstein, 1996.
[118] Smith, 1925, vol. II, p. 616.
[119] Smith, 1925, vol. II, p. 618.
[120] Smith, 1925, vol. II, p. 618.
[121] Smith, 1925, vol. II, p. 614–619.

Pope Gerbert had a unique training: monastic life from Christian teachers, pagan life from Latin classics, and academic learnings (astronomy and mathematics) from the Muslim teachers of tenth century Spain. His interests included literature, music, philosophy, theology, mathematics and the natural sciences. He became familiar with the Hindu numerals during his stay in Spain. Later, he wrote and taught in his own school in Rheims about the Hindu numerals with its positional notations and rules related to the arithmetical operations.[122] Gerbert used the nine signs of the Hindus to construct his abacus and "gave the multiplication or the division of each number, dividing and multiplying their infinite numbers with such quickness that, as for their multiplication, one could get the answer quicker than he could express in words."[123]

Gerbert wrote five books dealing with science and mathematics that are now lost. What we know about Gerbert is from the writings of his students, particularly Richer, the son of a French nobleman,[124] and through his letters.[125] Richer later became a monk in St. Rémy of Rheims and wrote a popular book on the history of France, *Historia Francorum.*[126]. Nikolai Bubnov, a Russian scholar, extensively studied the works of Gerbert and published a book about his contributions to mathematics, in Latin.[127] Recently, another book on the life of Gerbert was published by Pierre Riché in French.[128]

In one letter that was written to Constantine, monk of Fleury in 980 CE, Gerbert explained the Hindu numerals and mathematics to his monk friend using the abacus that was popular in Spain. Following are the excerpts of the letter: "Do not let any half-educated philosopher think they [Hindu numerals and mathematics] are contrary to any of the arts or to philosophy. . . For, who can say which are digits, which are articles, which the lesser numbers of divisors, if he disdains sitting at the feet of the ancient?"[129]

Richer described Gerbert's abacus in the following words: "in teaching geometry, indeed, he expended no less labor [than in teaching astronomy]. As an introduction to it he caused an abacus, that is, a board of suitable dimensions, to be made by a shield maker. Its length was divided into 27 parts [columns] on which he arranged the symbols, nine in number, signifying all numbers. Likewise, he had made out of horn a thousand characters, which indicated the multiplication or division of each number when shifted about in the 27 parts of the abacus. [He manipulated the characters] with such speed in dividing and multiplying large numbers that, in view of their very great size, they could be shown [seen] rather than be grasped mentally by words. If anyone desires to know this more thoroughly let him read his book which he

[122]Lattin, 1961.
[123]Darlington, 1947.
[124]Darlington, 1947.
[125]Lattin, 1961.
[126]Richer, 1964–67.
[127]Bubnov, 1899.
[128]Riché, 1987.
[129]Lattin, 1961, p. 45.

wrote to Constantine, the grammaticus; for there he will find these matters treated completely enough."[130]

Another disciple of Gerbert, Bernelinus, has described his abacus as consisting of a smooth board with a uniform layer of blue sand. For arithmetical purposes, the board was divided into 30 columns, of which 3 were reserved for fractions. These 27 columns were grouped with three columns in each group. These were marked as C (*cenlum*), D (*decem*), and S (*singularis*) and M (*monas*). Bernelinus gave the nine numerals and suggested that the Greek letters could also be used instead.[131]

Pope Sylvester lived in a period that is before the times of Fibonacci of Italy and Roger Bacon of England, both credited for the introduction of the Hindu numerals in Europe. Unfortunately, Pope Sylvester II died in 1003, less than four years after he was crowned as Pope. He was a champion of scientific scholarships along with his usual duties as a religious leader.

****

In summary, this chapter provides a brief sketch of the contributions of the Hindus, and reminds us of the mathematical tools that were perfected by the Hindus more than a thousand years ago and which have become a part of mainstream mathematics. "Think of our notation of numbers, brought to perfection by the Hindus, think of the Indian arithmetical operations nearly perfect as our own, think of their elegant algebraic methods, and then judge whether the Brahmins on the banks of the Ganges are not entitled to some credit," questions Cajori.[132] The answer is definitely in the affirmative.

[130]Lattin, 1961, p. 46. The 27 columns in his abacus is an interesting number. The ancient Hindus divided the ecliptic circle into 27 parts. (Read 4.2)

[131]Cajori, 1980, p. 116.

[132]Cajori, 1980, p. 97.

CHAPTER 4

Astronomy

As mentioned in Chapter 1, the American Association for the Advancement of Science (AAAS) ranked Hindus' contributions to astronomy among the top 100 scientific discoveries in human history. This recognition was a result of careful and systematic observations of the sky by the ancient Hindus. They noticed changes in the positions of some luminaries (planets, meteors, comets) against the fixed background of other luminaries (stars), tried to know the shape of the Earth, looked for an explanation for the various phases of the Moon, the changing seasons, designed their own luni-solar calendar, assigned motion to the earth in their cosmological model, and correctly assigned the age of the universe of the order of billions of years. Once scientific observations were made and conclusions were drawn, the ancient Hindus documented the facts into stories or poetic verses. For example, a story was written in which Rohiṇī (Aldebaran; a star in constellation Taurus), the celestial female deer, was pursued by the stars (male) of Orion, labeled as the celestial stag. Sirius in his role as a hunter pins Orion with his arrow to protect Aldebaran. The line that connects Sirius with Aldebaran go through the Belt of Orion, indicating the arrow's flight. This story is provided in the *Ṛgveda*.[1] Similarly, the Big Dipper or Ursa Major, with seven stars was labeled as *saptrishi mandal* (group of seven sages).

The Sanskrit word for astronomy is *khagola-śāstra*. Another term, *jyotiṣa-śāstra* (astrology) covers considerable astronomy too. Hindu astronomy partly flourished out of the need to perform religious rituals on proper days at particular times that were governed by the positions of the Sun or the Moon with respect to various constellations. The ancient Hindus performed rites at sunrise and sunset, at the rising and setting of the Moon, and at other well-defined entrances of the Moon or the Sun into particular constellations. These needs required keen observations and mapping of the sky. A special class of priests (*khagola-śāstri*, scholars of the sky) made observations of astronomical events, including the motions of the planets, and documented them in their hymns.[2] The *Yajurveda* mentions *Nakṣatra-daraśa*[3] as the term for astronomer. This term is made up of two words: *nakṣatra* means constellation or a prominent star and *daraśa* means seer or observer, signifying a person who studies astronomy. The *Chāndogya-Upaniṣad* mentions *nakṣatra-vidyā* (knowledge of constellations or astronomy) as a discipline. Most Hindu festivals are governed by astronomical events. For example, *Saṁkrānti*, an important festival in India, is

[1] Krupp, October 1996. Krupp is an internationally known astronomer and works at the Griffith Observatory in Los Angeles. Krupp failed to provide a proper reference in his article. I have not been able to trace it in the *Ṛgveda*. However, knowing the reputation of Krupp, I have decided to share it with the reader.

[2] see Brennard, 1988; Burgess, 1893; Paramhans, 1991; Saha and Lahri, 1992; Shukla and Sarma, 1976; Somayaji, 1971; Subbarayappa and Sarma, 1985.

[3] *Yajurveda*, 30: 10.

celebrated on the day when the Sun moves (apparent motion) away from one *rāśi* (zodiac) and enters in a new zodiac. There are 12 zodiac signs and, therefore, twelve *Saṁkrānti* festivals every year. The most popular of these festivals is *makar-Saṁkrānti*. This is the day when the Sun moves from the *dhanu* zodiac (Sagittarius) to *makar* (Capricorn) zodiac. On this day, the Hindus visit temples, fast for the day, donate money to needy people, feed hungry people, and some bathe in the holy rivers. This religious need required regular scientific observations of the constellations and the apparent motion of the Sun.

The ancient Hindus designed sophisticated instruments to facilitate their observations. Yukio Ōhashi, a Japanese scholar, based on his research, notice the following instruments that the Hindus had for astronomical observations during the period of Āryabhaṭa I:[4]

1. Chāyā-yantra (Shadow-Instrument)
2. Dhanur-yantra (semi-circle instrument)
3. Yaṣṭi-yantra (staff)
4. Cakra-yantra (circular instrument)
5. Chatra-yantra (umbrella instrument)
6. Toya-yantrāṇi (water instrument)
7. Ghaṭikā-yantra (clepsydra)
8. Kapāla-yantra (clepsydra)
9. Śaṅku-yantra (gnomon)

The *Taittirīya-Brāhmaṇa* advised the *khagola-śāstris* to the study the stars before sunrise to figure out the exact time of the rituals. "The position of an auspicious star [relative to the Sun] has to be determined at sunrise. But when the Sun rises, that star would not be visible. So, before the Sun rises, watch for the adjacent star. By performing the rite with due time adjustment, one would have performed the rite at the correct time."[5]

The ancient Hindus knew the shape of the Earth as spherical from the earliest periods. The *Śatapatha-Brāhmaṇa*, an ancient book of the Hindus, mentions the spherical shape of the earth: ". . . womb is spherical and moreover this terrestrial world doubtless is spherical in shape."[6] In his book *Geography*, Strabo (ca. 63 BCE - 25 CE), a Greek traveler and historian, mentions that the Indians, like the Greeks, believed in the spherical shape of the Earth: "According to the Brachmanes, the world . . . is of a spheroidal figure.[7]

[4]Ōhashi, 1994.
[5]*Taittirīya-Brāhmaṇa*, 1: 5: 2: 1.
[6]*Śatapatha-Brāhmaṇa*, 7: 1: 37.
[7]Strabo, 15: 1: 59.

Al-Bīrūnī (973–1050 CE) also affirms this view: "According to the religious traditions of Hindus, the Earth on which we live is round."[8] The key word in this quotation is "traditions." There is about a thousand years of time gap between Strabo and al-Bīrūnī. Since the Vedic period, the Earth was considered to be spherical. Al-Bīrūnī also quoted the Hindu astronomers to indicate that the size of the Earth was very small in comparison to the visible part of the universe:[9] "These are the words of Hindu astronomers regarding the globular shape of heaven and earth, and what is between them, and regarding the fact that the earth situated in the center of the globe, is only a small size in comparison with the visible part of heaven."

Āryabhaṭa I used an analogy of a *kadamba* flower (*neolamarckia cadamba*) (Figure 4.1) to demonstrate the distribution of various life forms on the Earth: "Half of the sphere of the Earth, the planets, and the asterisms is darkened by their shadows, and half, being turned toward the Sun, is lighted according to their size. The sphere of the earth, being quite round, situated in the center of space, in the middle of the circle of asterisms [constellations or stars], surrounded by the orbits of the planets, consists of Earth, water, fire, and air. Just as the ball formed by a *kadamba* flower is surrounded on all sides by blossoms just so the Earth is surrounded on all sides by all creatures terrestrial and aquatic."[10]

Figure 4.1: Kadamba flower (taken from Wikimedia).

There are numerous accounts in the Hindu literature indicating that the Moon gets its luminosity from the Sun. The *Ṛgveda* tells us that "He [the Moon] assumed the brilliancy of the Sun,"[11] "He [the Moon] is adorned with Sun's beams . . .,"[12] or, "the Sun-God Savitar bestows his sunlight to his Lord, the Moon."[13] The *Yajurveda* (Śukla) also tells us that "the Moon whose rays are the Sun's ray."[14]

Al-Bīrūnī (973–1050 CE), in stating the Hindu view that planets are illuminated by the Sun, wrote: "the Sun aloft is of fiery essence, self-shining, and 'per accidens' illuminates other

[8] Sachau, 1964, vol. 1, p. 233.
[9] Sachau, 1964, vol. 1, p. 269.
[10] *Āryabhaṭīya*, *Gola*, 5–7.
[11] *Ṛgveda*, 9: 71; 9.
[12] *Ṛgveda*, 9: 76: 4.
[13] *Ṛgveda*, 10: 85: 9; *Atharvaveda*, 14: 1: 9.
[14] *Yajurveda*(Śukla), 18: 40.

stars when they stand opposite to him."[15] "The Moon is watery in her essence," writes Al-Bīrūnī. "therefore the rays which fall on her [Moon] are reflected, as they reflect from the water and the mirror toward the wall."[16]

Varāhamihira stated that, in the words of al-Bīrūnī: "The Moon is always below the Sun, who throws his rays upon her, and lit up the one half of her body, whilst the other half remains dark and shadowy like a pot which you place in the sunshine. The one half which faces the Sun is lit up, whilst the other half which does not face it remains dark."[17] "If the Moon is in conjunction with the Sun [new moon], the white part of her turns toward the Sun, the black part toward us. Then the white part sinks downward toward us slowly, as the Sun marches away from the Moon."[18]

The above statements from al-Bīrūnī demonstrate that the ancient Hindus knew that the moonlight in the night sky is actually the reflected sunlight. The night sky does not have the Sun. If so, how could it illuminate the Moon? Based on the statements provided, the ancient Hindus had the understanding of the changing geometry of the Earth, the Moon, and the Sun. Information to support this argument is provided later, especially in relation to Rāhu and Ketu. The ancient Hindus knew that the Sun does not dissolve after the sunset. The Sun does not set or rise. (Section 4.1.) Al-Bīrūnī states that "every educated man among Hindu theologians, and much more so among their astronomers, believes indeed that the Moon is below the Sun, and even below all the planets."[19]

Āryabhaṭa I (Figure 4.2) explained the geometry of the solar system and the universe: "Half of the globe of the Earth, the planets, and the stars are dark due to their own shadows; the other halves facing the Sun are bright in proportion to their sizes."[20] "Beneath the stars are Saturn, Jupiter, Mars, the Sun, Venus, Mercury, and the Moon, and beneath these is the Earth . . ."[21] This observation of the heavenly objects, as given by Āryabhaṭa I, is correct for an observer on the Earth.

The apparent path of the Sun as viewed from the Earth against the background of stars is called the ecliptic. This path results from the Earth's motion around the Sun against the backdrop of the celestial sphere. The Moon with its changing phases moves around the Earth in a plane that is close to the ecliptic plane. The plane of the Moon lies 5° north or south of the ecliptic plane. The two points at which the Moon's path crosses the elliptic plane, known as the descending and ascending nodes. The nodal points are not fixed and move along the ecliptic plane of the Earth's motion. The ancient Hindus called them Rāhu and Ketu, respectively. The rotational periods of Rāhu and Ketu as defined in Hindu astrological charts, the horoscopes, are

[15] Sachau, 1964, vol. 2, p. 64.
[16] Sachau, 1964, vol. 2, p. 66.
[17] Sachau, 1964, vol.2, p. 66.
[18] Sachau, 1964, vol. 2, p. 67.
[19] Sachau, 1964, 2, p. 67.
[20] *Āryabhaṭīya*, *Gola*, 5.
[21] *Āryabhaṭīya*, *Kālakriyā*, 15.

Figure 4.2: Āryabhaṭa I statue on the premise of the Inter-University Center for Astronomy and Astrophysics, Pune, India (taken from Wikimedia)

the same as that of these nodal points. These two nodes are diametrically opposite to each other and so are the Rāhu and the Ketu in the horoscope.

These nodal points describe the relation of the Moon and the Sun to the Earth. Along with the individual effects of the Sun and Moon, there is also a collective effect exhibited by the Sun and the Moon, as noticed in lunar and solar eclipses. The solar eclipse and the lunar eclipse can be explained with the help of these nodal points. The ancient Hindus knew this fact and tried to explain eclipses by the motion of Rāhu and Ketu, the two nodal points.

The *Ṛgveda* mentions solar eclipse[22] which is a common observation in most cultures. An explanation for the cause was warranted. In Hindu scriptures, as given in the *Ādi-parva* of *Mahābhārata*, eclipses are described as the swallowing of the Sun or the Moon by two demons, Rāhu or Ketu. As the story goes, once the gods decided to churn the ocean to get the divine nectar (*amṛta*) to become immortal. As *amṛta* came out from the churning process, all gods started to drink it. A demon took the guise of a god and drank *amṛta* too. When he gulped it, the Sun and the Moon recognized him. They both warned Lord Viṣṇu about his transgression. Lord Viṣṇu acted promptly and slit his throat. By then, however, *amṛta* had already entered in his body and, therefore, he could not die. His head is called Rāhu and torso Ketu. Ever since there has been a long lasting feud of Rāhu and Ketu with the Sun and the Moon. They chase them across the sky and try to swallow them. Whenever Rāhu or Ketu succeed, an eclipse occurs. An eclipse is therefore symbolizes a momentary victory of Rāhu and Ketu over the Sun or the

[22] *Ṛgveda*, 5: 40: 5–9.

Moon. In essence, it is a temporary victory of evil over good. Therefore, eclipse is considered as an inauspicious occasion in Hindu scriptures. Hindus, therefore, pray during an eclipse and bathe in their holy rivers to purify themselves afterward.

The *Chāndogya-Upaniṣad* provided an account of the lunar eclipse: "From the dark, I go to the varicolored. From the varicolored, I go to the dark. Shaking off evil, as a horse his hairs; shaking off the body, as the Moon releases itself from the mouth of Rāhu."[23] Here "varicolored" is indicative of the corona rings that appear during an eclipse. The mention of Rāhu is indicative of the role of the nodal points.

The *Atharvaveda* provides the cause of a solar eclipse: "Peace to us during lunar eclipse, peace to us during the period when Rāhu swallow the Sun."[24] Āryabhaṭa I used scientific terms and explained the cause of solar and lunar eclipses as the Moon blocks the Sun or the Earth comes in between the Sun and the Moon. He also provided the method to calculate the area of the Moon or the Sun that would be affected during an eclipse: "The Moon obscures the Sun and the great shadow of the Earth obscures the Moon."[25] This is a clear indication that the geometry of eclipse was known to Āryabhaṭa I.

"When at the end of the true lunar month the Moon, being near the node, enters the Sun, or when at the end of the half-month the Moon enters the shadow of the Earth that is the middle of the eclipse which occurs sometimes before and sometimes after the exact end of the lunar month or half-month."[26]

"Multiply the distance of the Sun by the diameter of the Earth and divide (the product) by the difference between the diameters of the Sun and the Earth: the result is the length of the shadow of the Earth (i.e., the distance of the vertex of the Earth's shadow) from the diameter of the Earth (i.e., from the center of the Earth).[27]

Thus,

$$\text{Length of the Earth's Shadow} = \frac{\text{Sun's distance} \times \text{Earth's diameter}}{\text{Sun's diameter} - \text{Earth's diameter}}$$

"The difference between the length of the Earth's shadow and the distance of the Moon from the Earth multiplied by the diameter of the Earth and divided by the length of the Earth's shadow is the diameter of the Earth's shadow (in the orbit of the Moon)."[28]

Al-Bīrūnī confirms that the cause of solar and lunar eclipse was known to the Hindus. "It is perfectly known to the Hindu astronomers that the Moon is eclipsed by the shadow of the Earth, and the Sun is eclipsed by the Moon."[29]

[23] *Chāndogya-Upaniṣad*, 8: 13: 1.
[24] *Atharvaveda*, 19: 9: 10.
[25] *Āryabhaṭīya*, *Gola*, 37.
[26] *Āryabhaṭīya*, *Gola*, 38.
[27] *Āryabhaṭīya*, *Gola*, 39, taken from Shukla and Sarma, 1976, p. 152.
[28] *Āryabhaṭīya*, *Gola*, 40.
[29] Sachau, 1964, vol. 2, p. 107.

4.1 HELIOCENTRIC SOLAR SYSTEM

Most planetary models during the ancient period considered a geocentric system where the Earth remained stationary, like in the Ptolemy's model. Planets moved around the earth in epicyclic motions in these models. Āryabhaṭa I, on the contrary, came up with a detailed and innovative model of the solar system in which the Earth was in axial motion. With the known spherical shape of the Earth, Āryabhaṭa I assigned diurnal motion to the Earth, and was able to explain the repeated occurrence of day and night. He somehow knew the time difference between various locations on earth: "Sunrise at Lanka (Sri-Lanka) is sunset at Siddhapura, mid-day at Yavakoti (or Yamakoti, Indonesia), and mid-night at Romaka (Rome)."[30] With the current knowledge, we can safely say that Āryabhaṭa I somehow knew the longitudes of these locations and could correctly infer the time differences.

The above statement is remarkable since it involves information that requires simultaneous observation. Needless to say, Āryabhaṭa I had no way to make a phone call since the phones were not available. Similarly, all other means to contact with people in Rome was not possible for a person sitting in India. The only possibility is to predict it using geometry and astronomy in which the relative motion of the Sun and the Earth, including its shape, is known to the person. This is similar to what Eratosthenes did when he made simultaneous measurements at Syene and Alexandria, both in Egypt, to measure the size of the Earth.[31]

Several Hindu scriptures indicate that the Sun constantly illuminates the Earth. Sunrise or sunset happen depending on the side of the Earth illuminated by the Sun at that particular instant. Following are some of representative statements: "Actually the Sun neither rises nor sets. . . When the Sun becomes visible to people, to them He [the Sun] is said to rise; when He [the Sun] disappears from their view, that is called his [the Sun] setting. There is in truth neither rising or setting of Sun, for he is always there; and these terms merely imply his presence and his disappearance," suggests *Viṣṇu-Purāṇa*.[32]

"He [the Sun] never sets or rises. When [we] think that he is setting, he is only turning round, after reaching the end of the day, and makes night here and day below. Then, when [we] think he is rising in the morning, he is only turning round after reaching the end of the night, and makes day here and night below. Thus, he (the Sun) never sets at all," suggests the *Āitareya-Brāhmaṇa*.[33]

"Never did the Sun set there nor did it rise. . . the Sun neither rises nor sets. He who thus knows this secret of the *Vedas*, for him, there is perpetual day,"[34] suggests the *Chāndogya-Upaniṣad*.

[30] *Āryabhaṭīya*, *Gola*, 13. The locations of Siddhapura is not known. Assuming Āryabhaṭa I to be correct, this longitude falls somewhere in the continent of America, near Mexico.

[31] Brown and Kumar, 2011.

[32] *Viṣṇu-Purāṇa*, 2: 8.

[33] *Āitareya-Brāhmaṇa*, 14: 6; taken from Subbarayappa and Sarma, 1985, p. 28.

[34] *Chāndogya-Upaniṣad*, 3: 11: 1–3; taken from Subbarayappa and Sarma, 1985, p. 28. Some translators have translated the text somewhat differently. However, in most translations, "Sun never sets" is stated.

4.1.1 UJJAIN, GREENWICH OF THE ANCIENT WORLD

Sri Lanka was used as a reference point because the prime meridian of Ujjain, a city in India, (Longitude 75°43′E, Latitude 23°13′N) intersects the equator near Sri Lanka. The location of Ujjain played an important role in astronomy during the ancient and medieval periods. Ujjain was the Greenwich of the ancient and medieval world and most ancient astronomers in India and Arabia used this city as a reference point for their astronomical observations. Āryabhaṭa I defined the distance between Ujjain and Sri Lanka as one-sixteenth of the Earth's circumference in the north direction.[35] This gives the latitude of Ujjain as $\frac{360}{16} = 22°30'$ which is quite close to the actual number. A difference of 1° in latitude creates an error of 69 miles in distance. Therefore, the value given by Āryabhaṭa I for the latitude differs with the actual value by 43′ which is equivalent to less than 50 miles. People in most metropolitan cities would agree that this is a small distance. For example, suburban sites in New York or Los Angeles can be 50 miles away from the city center and still would be considered as a part of these cities. Since the settlements around Ujjain have evolved over the last thousand years, it may easily be the distance of modern city center and the ancient location of the observatory.

During the medieval period, Ujjain was considered as the prime meridian in the Middle East. More details on Ujjain is provided in Section 4.4.

4.1.2 DIURNAL MOTION OF THE EARTH

Āryabhaṭa I assigned diurnal motion to the Earth and kept the Sun stationary in his astronomical scheme. According to Āryabhaṭa I the motion of the stars that we observe in the sky is an illusion. To explain the apparent motion of the Sun, Āryabhaṭa I used an analogy of a boat in a river. "As a man in a boat going forward sees a stationary object moving backward just so in Sri-Lanka a man sees the stationary asterisms (stars) moving backward exactly toward the West."[36]

Āryabhaṭa I Āryabhaṭa I was the Head at the famous Nālandā University near modern Patna. He composed a book, *Āryabhaṭīya*, dealing with mathematics (*gaṇita-pada*), spherical astronomy (*gola-pada*), and time-reckoning (*kāla-kriyā-pada*). The book was composed in 499 CE by Āryabhaṭa I when he was only 23 years old. His work made a major impact in India for several centuries as the following writers like Brahmgupta (born around 598 CE) and Varāhamihira (505–587 CE) wrote extensive commentaries on his work.

Āryabhaṭīya is an invaluable document for the historians of science; it provides an account of the ancient sciences of the Hindus. The content of this book was not a new knowledge created by Āryabhaṭa I. He was emphatic to not take the credit and labeled the content of his book as "old knowledge."

[35] *Āryabhaṭīya*, *Gola*, 14; Shukla and Sarma, 1976, p. 123–126.

[36] *Āryabhaṭīya*, *Gola*, 9; to read more about Āryabhaṭa I, read Hooda and Kapur, 1997; and Shukla and Sarma, 1976.

Āryabhaṭīya has 118 metrical verses subdivided into four chapters. The first chapter, *Daśa-gītikā* has ten verses and provided astronomical constants. The second chapter is on mathematics. The third chapter is on time-reckoning, and the fourth chapter concerns spherical astronomy. In this book, Āryabhaṭa I provided the value of π as *approximately equal to* 3.1416, the solution of indeterminate equations and quadratic equations, theory of planetary motions, and calculations of the latitudes of planets. And most important of all, a millennium before Copernicus, he assigned axial motion to the Earth in his astronomical model and kept the stars stationary.

In Āryabhaṭa I's honor, the first artificial satellite of India was launched in 1975 and was named after him. The International Astronomical Union has also named a lunar crater after Āryabhaṭa I in 1979.

The interpretation is that a person standing on the equator, that rotates from the west to the east, would see the asterisms (constellations) moving in the westward motion. This clear grasp of Earth's rotational motion is splendidly explained in the analogy of a boat man by Āryabhaṭa I.

Interestingly, about one millennium after Ārybahaṭa I, Copernicus used a similar argument to assign motion to the Earth. "For when a ship is floating calmly along, the sailors see its motion mirrored in everything outside, while on the other hand they suppose that they are stationary, together with everything on board. In that same way, the motion of the earth can unquestionably produce the impression that the entire universe is rotating."[37] This similarity between Ārybahaṭa I's statement and Copernicus' statement is intriguing. Did Copernicus know of the work of Āryabhaṭa I? This issue is not clearly resolved as yet. However, there is a possibility of Copernicus knowing the work of Āryabhaṭa I, as discussed in Section 4.4.

******

Once the rotational (axial or spin) motion of the Earth is established, what kind of issues it creates to explain the observed phenomena? Why do we have different seasons? Does it lead to the heliocentric theory of solar system? Let us investigate further to answer these questions.

With spin motion assigned to the Earth, a set of other questions immediately arise: Is there a motion of the Sun? This question pops up since there is no longer a necessity to explain day and night with the Sun's motion. According to Āryabhaṭa I, the Earth spins on its axis like a merry-go-round. However, we do not experience any fly-away feeling on the Earth as we do on a merry-go-round. The spin motion of the Earth also creates problems to explain flights of flying birds. How do they go back to their nest with the Earth spinning so fast, especially if they fly to the West? Assigning any motion to the Earth seems contrary to human experiences. It is a much bigger triumph to assign any kind of motion to the Earth than to just add orbital motion to the already known spinning (rotational) motion of the Earth.

[37] Copernicus' *On the Revolution*, Book 1, Chapter 8; Copernicus, 1978, p. 16.

This leads to the next question. If day and night are due to the Earth spinning in one place, why do we have different seasons? Why do we observe northward or southward motions of the Sun? Āryabhaṭa I considered constellations and stars to be stationary in the sky and attributed their apparent motion to the moving Earth. Can the Sun be also stationary like other stars? No where did Āryabhaṭa I struggle in dealing with such questions in his book. His statements are fairly conclusive and straight forward. In summary, the axial rotation of the Earth complicates the simplicity of the geocentric system. On the contrary, in a heliocentric system, the axial rotation is a necessity.

In describing the spin motion of the Earth, Āryabhaṭa I makes another explicit statement, "The revolutions of the Moon multiplied by 12 zodiac are signs [*rāśi*]. The signs multiplied by 30 are degrees [360°]. The degrees multiplied by 60 are minutes. . . . The Earth moves one minute in a *prāṇa*."[38]

Āryabhaṭa I provided the following definition of *prāṇa*, a unit of time: "One year consists of twelve months. A month consists of thirty days. A day consists of sixty *nāḍī*. A *nāḍī* consists of sixty *vināḍikā*. Sixty long syllables or six *prāṇas* make a sidereal *vināḍikā*. This is the division of time."[39]

Let us transcribe it in a modern set of standards. Assuming a day to be 24 hours long with a 360° rotation (86,400′), one *nāḍī* comes out to be 1,440′, *vināḍikā* equals to 24′, and *prāṇa* comes out to be four seconds. Therefore, Āryabhaṭa I's statement can be modified as follows: "The Earth rotates by an angle of one minute (1′) in 4 seconds." One minute in angle multiplied by 21,600 gives us 360°. Thus, 4 seconds multiplied by 21,600 should give us the time that is equal to one day (or 24 hours). This is the case when we multiply the two numbers and change the units to hours. Therefore, not only Āryabhaṭa I assigned spin motion to the Sun, he also correctly provided the speed of the spin. Āryabhaṭa I makes explicit statement elsewhere: "The Earth rotates through [an angle of] one minute of arc in one *prāṇa*."[40]

Āryabhaṭa I provided the periods of revolution for different planets, the Moon, and the Sun in one *yuga*: "In a yuga the revolutions of the Sun are 4,320,000, of the Moon 57,753,336, of the Earth 1,582,237,500, of Saturn 146,564, of Jupiter 364,224, of Mars 2,296,824, of Mercury and Venus the same as those of the Sun. of the Moon's apogee, 4,88,219; of [the *śighrocca* (conjunction)] of Mercury, 1,79,37,020; of (the *śighrocca*) of Venus, 70,22,388; of (the *śighrocca*) of the other planets, the same as those of the Sun; of the Moon's ascending node in the opposite direction (i.e., westward), 2,32,226. These revolutions commenced at the beginning of the sign Aries on Wednesday at sunrise at Sri Lanka (when it was the commencement of the current yuga.)"[41]. *Yuga* is an important concept in Hindu cosmology, as given in the Hindu scriptures (Section 4.3).

[38] *Āryabhaṭīya, Dasgītika*, 4.
[39] *Āryabhaṭīya, Kālakriyā*, 2.
[40] *Āryabhaṭīya, Dasgītika*, 6.
[41] *Āryabhaṭīya, Dasgītika*, 3–4

The "Moon's apogee" defines the point of the Moon's orbit when it is farthest from the Earth. The *śighrocca* of a planet is the imaginary body which is supposed to revolve around the Earth with the heliocentric mean velocity of the planet. Shukla and Sarma have done a careful analysis of these periods in their translation of *Āryabhaṭīya*, calculated the sidereal period in terms of days, and compared them with the modern values.[42] (Table 4.1.)

Table 4.1: Mean motion of the planets

Planet	Revolutions in 4,320,000 Years	Sidereal Period Āryabhaṭa I	Sidereal Period Moderns
Sun	4,320,000	365.25868	365.25636
Moon	57,753,336	27.32167	27.32166
Moon's apogee	488,219	3,231.98708	3,232.37543
Moon's asc. node	232,226	6,794.74951	6,793.39108
Mars	2,296,824	686.99974	686.9797
Śighrocca of Mercury	17,937,020	87.96988	87.9693
Śighrocca of Venus	7,022,388	224.69814	224.7008
Jupiter	364,224	4,332.27217	4,332.5887
Saturn	146,564	10,766.06465	10,759.201

As one can notice, all values are quite close to the modern accepted values. Interestingly, the period of Mercury and Venus are explicitly considered equal to the Sun by Āryabhaṭa I. Since Mercury and Venus are the only two planets inside the Earth's orbit, their orbital period, as observed from the Earth, comes close to that of the Sun. Obviously, this observation is in a geocentric system. Was it because Āryabhaṭa I believed in the geocentric solar system or because it was a prevalent practice of the period to describe motion as viewed from the Earth? In the very next verse, Āryabhaṭa I erased any doubt about the two different motions assigned to Mercury and Venus. Thus, the sidereal period of Mercury and Venus comes out to 87.97 and 224.70 days, in the calculation of Shukla and Sarma.[43] This compares well with the modern values of 87.97 and 224.70, respectively for Mercury and Venus. Thus, Āryabhaṭa I provides one statement in the geocentric system while the other in the heliocentric system with the Earth as the point of observation. This has intrigued astronomers through the ages.

[42] Shukla and Sarma, 1976, p. 7.
[43] Shukla and Sarma, 1976, p. 7.

According to a detailed analysis given by B. L. van der Waerden,[44] the motion of Mercury and Venus as given by Āryabhaṭa I were in a heliocentric model.[45] Van der Waerden makes the following assertions to back up his conjecture that Āryabhaṭa I proposed a heliocentric model and not a geocentric model:[46]

1. In a geocentric system, there is no need to assume axial rotation for the Earth as most observations, though wrong, are easier to explain. However, in a heliocentric system, we are forced to think of axial rotation. Āryabhaṭa I clearly argue for the axial rotation of the Earth and provides accurate period for the axial motion.

2. In the Midnight system of Āryabhaṭa I, the apogees (farthest point from the earth) of the Sun and Venus are both at 80°, and their eccentricities (a measure of astronomical orbit deviation from circularity) are also equal. This fact can be explained by assuming that the system was originally derived from a heliocentric system.[47] "In genuine epicycle theory for Venus, the eccenter carrying the epicycle is independent of the Sun's orbit. Its apogee and eccentricity are determined from observations of Venus, whereas the apogee and eccentricity of the Sun are determined from eclipse observations. The probability that the apogee and eccentricity of Venus coincide with those of the Sun is very small."[48]

3. The periods of revolution for the outer planets are essentially the same in the geocentric and heliocentric models. It is the planetary periods of the inner planets, Mercury and Venus, that separates the two theories. In a geocentric theory, these two periods will essentially be the same as the solar period. However, in a heliocentric system, these periods are quite different. Āryabhaṭa did assign different periods for the Sun, Mercury and Venus, indicating a heliocentric system.

4. The revolutions of Mercury and Venus considered by Āryabhaṭa I are heliocentric revolutions, not geocentric. These periods provided by Āryabhaṭa I have "no importance whatsoever" in a geocentric system.[49]

Based on the descriptions given in *Āryabhaṭīya*, Van der Waerden concludes that "it is highly probable that the system of Āryabhaṭa [I] was derived from a heliocentric theory by setting the center of the Earth at rest."[50] The reason for this kind of torturous path in the work

[44] B. L. van der Waerden (1903–1996) was a prolific author with several books on algebra, geometry, and astronomy to his credit. He taught at the University of Leipzig in Germany, the University of Amsterdam in the Netherlands, and the University of Zurich in Switzerland.

[45] van der Waerden, The Heliocentric System in Greek, Persian, and Hindu Astronomy, in the book by King and Saliba, 1987, p. 534.

[46] van der Waerden, The Heliocentric System in Greek, Persian, and Hindu Astronomy, in the book by King and Saliba, 1987, p. 530–535.

[47] van der Waerden, in the book by King and Saliba, 1987, p. 532.

[48] van Waerden, in the book by King and Saliba, 1987, p. 531.

[49] Van Waerden in the book by King and Saliba, 1987, p. 534.

[50] Van der Waerden in the book by King and Saliba, 1987, p. 534.

of Āryabhaṭa I is perhaps due to an overwhelming tendency among all early astronomers and their students, in the words of Van der Waerden, "to get away from the idea of a motion of the Earth."[51] In the history of astronomy, for convenience purposes, astronomers did transform the heliocentric theory into equivalent geocentric one. This was done by Tycho Brahe when he transformed Copernican heliocentric model into a geocentric one.[52]

Van der Waerden's conclusion that Āryabhaṭa I proposed a heliocentric model of the solar system has received independent support from other astronomers.[53] For example, Hugh Thurston came up with a similar conclusion in his independent analysis. "Not only did Āryabhaṭa believe that the Earth rotates, but there are glimmerings in his system (and other similar Indian systems) of a possible underlying theory in which the earth (and the planets) orbits the Sun, rather than the Sun orbiting the earth."[54] The evidence used by Thurston is in the periods of the outer planets and the inner planets. Āryabhaṭa basic planetary periods are relative to the Sun which is not so significant for the outer planets. However, it is quite important for the inner planets (Mercury and Venus).

The motion that Āryabhaṭa I assigned to the Earth is not a mere speculation of modern astronomers. Āryabhaṭa I's thesis was well known in the Middle East even after six centuries. Al-Bīrūnī (973–1050 CE) erroneously criticized Hindu astronomers for assigning motion to the Earth (Figure 4.3). He referred to the work of Varāhamihira, a Hindu astronomer, to support his idea of the geocentric universe: "If that were the case, a bird would not return to its nest as soon as it had flown away from it toward the west,"[55] and "stones and trees would fall."[56] In the first drawing, Figure 4.3, the bird flies to the West and leave his nest. After a while, when he comes back, the nest has moved considerably due to the motion of the Earth. This was the argument of al-Bīrūnī against the moving Earth. A similar argument was used by Aristotle (384–322 BCE) to favor his theory of the geocentric universe. Al-Bīrūnī's criticism validates the work of Āryabhaṭa I in India and its existence in the Middle East prior to the eleventh century.

After viewing various possibilities of the motion of the Earth, al-Bīrūnī favored the stationary Earth: "The most prominent of both modern and ancient astronomers have deeply studied the question of the moving of the Earth, and tried to refute it. We, too, have composed a book on the subject called *Miftah-ilm-alhai'a* (*Key of astronomy*), in which we think that we have surpassed our predecessors, if not in the words, at all events in the matter."[57] This depicts that at least some six centuries after the heliocentric theory was proposed by Āryabhaṭa I, the Islamic philosophers from Arabia still could not grasp the idea of moving Earth.

[51]Van der Waerden in the book by King and Saliba, 1987, p. 530.
[52]Van der Waerden in the book by King and Saliba, 1987, p. 530.
[53]Billard, 1977; Thurston, 1994 and 2000.
[54]Thurston, 1994, p. 188.
[55]Sachau, 1964, vol. 1, p. 276.
[56]Sachau, 1964, vol. 1, p. 277.
[57]Sachau, 1964, vol. 1, p. 277.

Figure 4.3: Al-Bīrūnī's argument against Āryabhaṭa's assignment of the motion of the earth. (Designed with the help of David Valentino)

Science texts still do not cover Āryabhaṭa's work along with the work of Copernicus. The opponents of Āryabhaṭa's heliocentric system are totally silent in providing explanation to why Āryabhaṭa assigned spin motion to the earth and not struggled to explain the simple astronomical observations with this spin motion. Also, why the periods of Mercury and Venus are not equal to the period of the Sun, as it should be in a geocentric system. Obviously, a lot more research on *Āryabhaṭīya* is needed to resolve this issue.

4.2 HINDU CALENDAR

Pañcāṅga (*Pañcā* = five and *aṅga* = limb, meaning five-limbed) is the term used for the Hindu almanac. The five limbs are: day (*vāra*), date (*tithi*), *nakṣatra* (asterism), *yuga* (period), and *kāraṇā* (half of *tithi*). *Pañcāṅga* is popular even today among the Hindus. Hindu priests use it for predicting eclipses, defining time of various rituals, including marriage, casting horoscopes, and for solemn entrance into a house (*grah-praveśa*) or business. Hindu families commonly use *Pañcāṅga* to check the day of fasting, auspicious times for worshiping, and days of festivals.

Most cultures have calendars that are either based on the motion of the Moon or the Sun. The regular appearance of the new or the full moon forms a basis of most lunar calendars, like the Islamic calendar. Solar calendars are based on the cyclic motion of the Sun in different zodiacs that is due to the orbital motion of the Earth around the Sun. The Hindu calendar is luni-solar in which the months are based on the motion of the Moon while the year is defined by the Sun. A year is the time the Earth takes to complete one revolution around the Sun, starting from *Meṣa* (Aries). This calendar is similar to the Jewish or Babylonian calendar that are also luni-solar.

In the Hindu Calendar, a month is divided into two equal parts, known as *pakṣa*, each of roughly fifteen days depicting the waxing and waning of the Moon. The *pakṣa* starting from the new moon to the full moon is considered the bright-half (*Śukla-pakṣa*) while the second part starting from a full moon to a new moon, is known as the dark-half (*Kṛṣṇa-pakṣa*).[58] The new moon day, when the longitude of the Sun and Moon are equal, is called *amāvāsya*. The full moon night, when the Sun and the Moon are 180° out of phase, is known as *pūrṇimā*. It gives a mean lunar year to be 354 days 8 hours 48 minutes and 34 seconds.

A day (*vāra*) begins at sunrise. The date (*tithi*) is indicative of the position of the Moon relative to the Sun. A month (*māsa*) starts and ends on *amāvāsya* (meaning dwelling together, implying conjunction of the Sun and the Moon, the new moon). The word *amāvāsya* is used in *Atharvaveda*[59] which signifies that the ancient Hindus knew the cause of the new moon during the Vedic period. The days in between *amāvāsya* and *pūrṇimā* are counted as numbers: *ekādaśī* (eleventh day of the fortnight), *caturthī* (fourth day of the lunar fortnight), etc. The ecliptic circle was divided into 27 parts, each consisting of 13°20′, called *nakṣatra*.

To understand the features of the Hindu calendar, let us compare it with the Western calendar that is popular internationally. The Western calendar, also called the Gregorian calendar, was proposed by Pope Gregory XIII in 1582. It was a modified form of the calendar established by Julius Caesar, known as the Julian calendar. This calendar was based on the Egyptian calendar of the period. The Catholic kingdoms adopted the Gregorian calendar soon after its inception. However, England resisted its use and adopted it in 1752, under some resistance from the Protestant majority.

The Western calendar is irregular and inconvenient to use because:

1. There exists no easy way to figure out the date of a particular day from simple observations.

2. Different months have different numbers of days. This creates difficulties in the business world where, at times, monetary transactions are made based on the day devoted to a particular task.

3. Because the span of a month is different for different months, performance records are difficult to compare.

4. There is a problem of the leap year. One has to remember the year to decide the number of days in the month of February. There is no possible way to figure it out using astronomical observations. One has to remember the empirically defined rules to figure this out.

Most people use the Western calendar since their childhood and are not familiar with alternatives, they do not realize its weaknesses. In the lunar calendar, one year equals 354 days

[58] see *Arthaśāstra*, 108.

[59] defined in *Atharvaveda*, 7: 79.

and, in the solar calendar, one year is roughly equal to 365 days. The difference of 11 days in a year can cause radically different seasons for the same month in two years in a lunar calendar that are about 15–17 years apart from each other. This is the case with the Islamic calendar.

Āryabhaṭa I explained the civil and sidereal days: "The revolutions of the Sun are solar years. The conjunctions of the Sun and the Moon are lunar months. The conjunctions of the Sun and Earth are [civil] days. The rotations of the Earth are sidereal days."[60]

This defines the sidereal day as the period from one star-rise to the next, civil days as one sun-rise to the next, and the lunar month, or synodic month, as from one new moon to the next new moon. The ancient Hindus, who knew both the lunar and solar calendars, realized that 62 solar months are equal to 64 lunar months. Therefore, they added one extra month after every 30–35 months.

The *Ṛgveda* described the Moon as "the maker of months" (*māsa-kṛt).*[61] "True to his holy law, he knows the twelve Moons with their progeny: He knows the Moon of later birth,"[62]. Here "the twelve Moons with their progeny" means the twelve months and "the Moon of the later birth" means the 13th month, the supplementary or the intercalary month of the luni-solar calendar. This is a clear indication of the luni-solar calendar during the *Ṛgvedic*-period.

The *Atharvaveda* also mentions the 13th month in some years. "He [Sun] who meets out the thirteenth month, constructed with days and nights, containing thirty members, . ."[63] The creation of 13th month or the intercalary month of thirty days is ascribed to the Sun, the Moon being the originator of the ordinary months of the year. This is a clear indication that the thirteenth month was added to keep up with the seasons since it is ascribed to the Sun.

Al-Bīrūnī explained the Hindu luni-solar calendar in his book, *Alberuni's India*: "The months of the Hindus are lunar, their years solar; therefore their new year's day must in each solar year fall by so much earlier as the lunar year is shorter than the solar (roughly speaking, by eleven days). If his precession makes up one complete month, they act in the same way as the Jews, who make the year a leap year of thirteen months . . . The Hindus call the year in which a month is repeated in the common language *malamasa* [*malamāsa*]. *Mala* means the dirt that clings to the hand. As such dirt is thrown away out of the calculation, and the number of months of a year remains twelve. However, in the literature, the leap month is called *adhimāsa*."[64] Kauṭilya (ca. 300 BCE), in his *Arthaśāstra*, mentions a separate intercalary month and calls it *malamāsa*.[65] This month is generally added every third year.[66]

In the Hindu calendar, most of the festivals have religious, social, and seasonal importance. In societies where the lunar calendar is in practice, the seasonal festivals are not much celebrated. India is an agricultural country where approximately 65% of the population still lives in villages.

[60] *Āryabhaṭīya, Kālakriyā*, 5.
[61] *Ṛgveda*, 1: 105: 18.
[62] *Ṛgveda*, 1: 25: 8.
[63] *Atharvaveda*, 13: 3: 8.
[64] Sachau, 1964, vol. 2, p. 20.
[65] *Arthśāstra*, 60; Shamasastry, 1960, p. 59.
[66] *Arthaśāstra*, 109; Shamasastry, 1960, p. 121.

During the *Holī* festival, a big fire is burned every year to symbolize the death of Holikā, the aunt of Lord Dhruva. People bring a sample of their harvest, and place it over a fire to roast the wheat or barley seeds which they tie to sugarcane. They share these seeds and sugarcane with friends and family members and decide whether the crop of wheat, barley and cane sugar is ready for harvest or not. During *Daśaharā*, the quality of barley and wheat seeds is tested in a social gathering; people carry sprouted seeds and share them with their friends. Similarly, after the monsoon season from July to September, one needs to get ready for the winter in India. Cleaning spider webs, dusting rooms, painting walls, and decorating the houses are common chores before *Dīpāvalī* (Dīvālī).

Most Hindu festivals are defined either by the position of the Moon or the Sun. *Makara-Saṁkrānti* (the Sun enters the sign of Makara (Capricorn) constellation in its northward journey), *Gaṇeśa-caturthī* (fourth day of the Moon, starting from *amāvāsya*, the new moon), *Kṛṣṇa-Janmāṣṭamī* (eighth day of the Moon) are all defined by the phase of the Moon or the Sun in a particular constellation. *Basant-Pañcamī* (fifth day of the new Moon), *Rām-Navamī* (a day to honor Lord Rāma, falls on the ninth day of the new moon), *Guru-pūrṇimā* (a day to honor teachers, always fall on the full moon), and *Nāga-Pañcamī* (a day to honor snakes, falls on the fifth day of the new moon) are some of these festivals that are defined by the Moon.

The Moon is seen in the sky on almost all nights unless it is close to the Sun. The position of the Sun can be fixed against a constellation only a little before sunrise or after sunset—the time when the sunlight is too weak to suppress the light of other stars. Hindu astronomy, unlike Western astronomy, mapped the sky with the phases of the Moon rather than with the stars. It simplified their calculations—at full moon, the position of the Sun can automatically be given by that of the Moon. Similarly, the position of the Sun can easily be determined with the different phases of the Moon.

The following are the months in the Hindu calendar:

1. *Caitra* (March-April)
2. *Vaiṣākha* (April-May)
3. *Jyaiṣṭha* (May-June)
4. *Āṣāḍha* (June-July)
5. *Śrāvaṇa* (July-August)
6. *Bhādrapad* (August-September)
7. *Āśvina* or *Kwār* (September-October)
8. *Kārttika* (October-November)
9. *Agarhayana* or *Aghan* (November-December)

10. *Pauṣa* (December-January)

11. *Māgha* (January-February)

12. *Phālguna* (February-March)

The names of the months are derived from the names of the *nakṣatra* (star or constellation) in which the Sun dwells (or nearby). Xuan Zang (or Hiuen Tsang), a Chinese traveler who visited India during the seventh century, and al-Bīrūnī who traveled in India during the eleventh century also used the similar names for various months in India.[67]

Nothing was more natural for the sake of counting days, months, or seasons than to observe the twenty-seven places which the Moon occupied in her passage from any point in the sky back to the same point. The location of the Moon and its shape provided the *tithī* (date) as well as the particular time in the night sky for astute observers. This procedure was considerably easier than determining the Sun's position either from day to day or from month to month. As the stars are not visible during daytime and barely visible at sunrise and sunset, the motion of the Sun in conjunction with certain stars was not an easily observable task. On the contrary, any Vedic shepherd was able to decide day and time easily with the observation of the Moon.

The ancient Hindus formulated a theory of creation which was cyclic in nature. The universe followed a cycle of manifestated and non-manifestated existence. A new unit of time, *yuga*, was chosen to define the period of this cycle. The *yuga* system is based on astronomical considerations, and is frequently mentioned and explained in the *Purāṇic* literature.

In the *yuga* system of the Hindus, a *mahā-yuga* (*mahā* means big in the Sanskrit language) is divided into four *yugas*: *Satya* or *Kṛta*, *Tretā*, *Dvāpar*, and *Kali*. The spans of *Satya-yuga*, *Tretā-yuga*, *Dvāpara-yuga*, and *Kali-yuga* are in the ratio 4: 3: 2: 1. According to the Hindu scriptures, life on the Earth diminishes in goodness as we go from Satya- to *Kali-yuga*. At present, we live in the age of *Kali-yuga* that started in 3102 BCE of the Julian Calendar.

A *Kali-yuga* is equal to 432,000 solar years. A *maha-yuga* is equal to 4 + 3 + 2 + 1 = 10 *Kali-yuga*, equivalent to 4,320,000 solar years. Seventy *maha-yuga* constitute a *manvantara* and fourteen *manvantara* constitute one *kalpa*. A *kalpa* is the duration of the day of Brahmā, the creator of the universe. The night is equally long and the creation dissolves into the unmanifested form during the night of Brahmā. After that, it starts again.[68] Thus, the total cycle of creation is about 8.5 billion years. It is this number that caught the attention of Carl Sagan, after recognizing this number to be so close to number accepted by modern science which is also of the order of billion of years. In contrast, just two centuries ago, a large number of scientists or scholars in the West considered earth to be just six thousand years old.

[67]For Hiuen Tsang, see Beal, book II, p. 72; for al-Bīrūnī, see Sachau, vol. 1, p. 217.
[68]Garuḍa-*Purāṇa*, Chapter 233.

4.3 HINDU COSMOLOGY

The *Ṛgveda* raises questions about the creation: "What was the tree, what wood, in sooth, produced it, from which they fashioned forth the Earth and Heaven?"[69] Or, "What was the place whereon he took his station? What was it that supported him? How was it? Whence Visvakarman [God], seeing all, producing the Earth, with mighty power disclosed the heaven."[70]

In Hindu theory of creation, matter was not created from nothing, as the *Bhāgavad-Gītā* (2: 16) tell us: "Nothing of non-being comes to be, nor does being cease to exist." The *Ṛgveda* gives an account of creation and explains the state of the universe just before the blast: "Then was neither non-existent nor existent: there was no realm of air, no sky beyond it. What covered in, and where? And what gave shelter? Was water there, unfathomed depth of water? Death was not then, nor was there aught immortal: no sign was there, the day's and night's divider. That one thing, breathless, breathed by its own nature: apart from it was nothing whatsoever. Darkness there was: at first concealed in darkness this all was indiscriminate chaos. All that existed then was void and formless: by the great power of *warmth* was born that unit."[71]

The *Ṛgveda* uses the analogy of a blast furnace of a blacksmith to explain the creation process: "These heavenly bodies produced with blast and smelting, like a smith. Existence, in an earlier age of Gods, from non-existence sprang. Thereafter were the regions born. This sprang from the productive power."[72] This is quite similar to the Big-Bang Theory in which the universe was created with a Big Bang, like a bomb explosion.

The *Srimad-Bhāgvatam* provides a description of the process of creation in which the whole universe was in noumenal form and came to existence under the desire of the Creator (God). The existence will continue for a period of time before the universe would again go back to its noumenal form of matter. This will form a cyclic (oscillating) process will continue forever.[73] The concept of matter that cannot be experienced was not a part of science only 100 years ago. Today, the issues of dark matter and energy as a form of matter are scientific realities. The *Bhagavad-Gītā* describes the creation of the universe as a transformation of noumenal matter into matter: "From the noumenal all the matter sprung at the coming of the day; at this coming of the night they dissolve in just that called the noumenal."[74]

For the Hindus, the end of each creation comes with heat death, somewhat similar to the accelerated global warming which is causing concerns to modern scientists. This process of destruction is described in the *Viṣṇu-Purāṇa*. "The first, the waters swallow up the property of Earth, which is the rudiment of smell; and Earth, deprived of its property, proceeds to destruction. Devoid of the rudiment of odor, the Earth becomes one with water. The waters then being much augmented, roaring, and rushing along, fill up all space, whether agitated or still. When

[69] *Ṛgveda*, 10: 31: 7
[70] *Ṛgveda*, 10: 81: 2.
[71] *Ṛgveda*, 10: 129: 1–4.
[72] *Ṛgveda*, 10: 7: 2, 3.
[73] *Srimad-Bhāgvatam*, 3, 11–12.
[74] *Bhāgavad-Gītā*, 8: 8.

the universe is thus pervaded by the waves of the watery element, its rudimental flavor is licked up by the element of fire, and, in consequence of the destruction of the rudiments, the waters themselves are destroyed. Deprived of the essential rudiment of flavor, they become one with fire, and the universe is therefore entirely filled with flame, which drinks up the water on every side, and gradually overspreads the whole of the world. While space is enveloped in flame, above, and all around, the element of wind seizes upon the rudimental property, or form, which is the cause of light; and that being withdrawn, all becomes of the nature of air."[75]

Al-Bīrūnī used the views of Varāhamihira, a fifth century Hindu philosopher, to explain the Hindu view of creation. "It has been said in the ancient books that the first primeval thing was darkness, which is not identical with the black color, but a kind of non-existence like the state of a sleeping person."[76] "Therefore, they [Hindu] do not, by the word creation, understand a formation of something out of nothing."[77] "By such a creation, not one piece of clay comes into existence which did not exist before, and by such a destruction not one piece of clay which exists ceases to exist. It is quite impossible that the Hindus should have the notion of a creation as long as they believe that matter existed from all eternity."[78]

Let us sum up the Hindus' theory of creation as provided above: There was a void in the beginning. This void was not the one that we perceive in a strict physical sense; this void was full of energy, in analogy, similar to the fields of modern physicists. The voluminous writings in Hindu literature does not describe the special creation as arising out of nothing. Almost all the scriptures of the Hindus, including the earliest *Ṛgveda*, advocate that the present form of the universe evolved from the noumenal form of matter.

"Void" or "nonexistence" is like the noumenal matter or dark matter proposed by the modern scientists and philosophers; like the wavefunctions of Schrödinger to explain the microscopic reality that cannot be experienced but, when squared, gives the probability of existence of a particle. The noumenal matter is beyond the senses' experience but gives rise to a manifested form of matter with a blast under the desire of God.[79] The language used to describe the process of creation in the *Vedas* is at least four thousand years old. The account of the Hindu theory of creation by al-Bīrūnī is about 1,000 years old. Yet, there are striking similarities in the modern theory of creation and the Hindu theory of creation. In both theories, the present universe was created with a blast. In addition, the ancient Hindus believed that the creation process is cyclic in nature, *i.e.* it goes through the cycle of creation and destruction. The destruction will start with

[75] *Viṣṇu-Purāṇa*, 6: 4.
[76] Sachau, 1964, vol. 1, p. 320.
[77] Sachau, 1964, 1, p. 321.
[78] Sachau, 1964, vol. 1, p. 323.
[79] Kant defined the term, *noumenon*, which means a thing that cannot be perceived and can only be inferred from experience. It is a product of intellectual intuition, like the interaction of electrical fields that gives rise to a force on a charge particle when placed near other charge particles. Such intuitions are an integral part of most religions as well science where realities are defined from inferred experiences. The analogy mentioned above is an effort to describe *nonexistence* of the ancient Hindus by sharing similar concepts in physics. However, we are dealing with two different domains of knowledge and the nonexistence of the ancient Hindus is not the dark matter or Schrödinger's wavefunction.

an increase in heat that will give rise to an increase in the water levels of the oceans. Eventually, the heat will become so high that it will destroy all life forms.

Since everything animated in the universe should have a cause or the beginning, what was the beginning of the void or the beginning of the noumenal matter? Instead of drawing a picture of the beginning of the universe, the Hindus exposed the limit of human inquiry. The *Ṛgveda* says: "Who verily knows and who can here declare it, whence it was born and whence comes this creation? The Gods are later than this world's production. Who knows then whence it first came into being? He, the first origin of this creation, whether he formed it all or did not form it, Whose eye controls this world in highest heaven, He verily knows it, or perhaps *He knows not*."[80] In this way, the *Ṛgveda* exposes the limits of human inquiries to define the first cause of creation.

4.4 DIFFUSION OF HINDU ASTRONOMY

The sacred books of the Hindus and Āryabhaṭa I's and Brahmgupta's books on astronomy reached the Middle East and China, and caused a spurt of new books on astronomy in these regions. With the influence of the Middle East in Europe, it also indirectly impacted there too. Several noted European astronomers, including Copernicus and Kepler, read translations of the books from the Middle East that were based on Hindu astronomy.

4.4.1 THE MIDDLE EAST

If we had to choose just ten astronomy books in human history, *Zīj al-Sindhind* of al-Khwārizmī (c. 800–850 CE) would be among these books. Al-Khwārizmī, Abu Ja'far Muḥammad ibn Mūsā (ca. 800–847 CE) was a member of al-Ma'mun's *Bayt al-Hikma*. The motion of seven celestial bodies, the mean motions, and the positions of apogee and the nodes are described in his *Zīj al-Sindhind*.[81] As the title indicates, this book was based on Hindu astronomy. The contents and the mathematical procedures[82] used by al-Khwārizmī in his book agree well with *Brahmsphuta-siddhanta*(composed in 628 CE) of Brahmgupta, a Hindu astronomer.[83]

"al-Khwārizmī's . . . treatise on astronomy was . . . a set of tables concerning the movements of the Sun, the Moon and the five known planets, introduced by an explanation of its practical use. Most of the parameters adopted are of Indian origin, and so are the methods of calculation described, including in particular use of the sine," concludes Régis Morelon in his analysis.[84] al-Khwārizmī's book has been translated into English by Neugebauer.[85] This book of al-Khwārizmī

[80] *Ṛgveda*, X: 129: 6–7; for more information on Hindu cosmology, see Jain, 1975 and Miller, 1985.

[81] Translated by Neugebauer, 1962.

[82] Goldstein, 1967; Kennedy and Ukashah, 1969.

[83] Salem and Kumar, 1991, p. 47.

[84] Régis Morelon, Eastern Arabic Astronomy Between the Eighth and the Eleventh Centuries, in the book by Roshdi Rashed, 1996, vol. 1, p. 21. He taught in CNRS in Paris for many years and later served as director of IDEO (Dominican Institute of Oriental Studies) in Cairo from 1984 to 2008.

[85] Neugebauer, 1962.

became one of the most documented books of astronomy in Europe during the medieval period. The famous Toledo Tables and the Alfonsine Tables were based on this book. In the title of his book, al-Khwārizmī has acknowledge Hindus' contribution to astronomy. Several medieval and modern historians have written about the connection of *Zīj al-Sindhind* to Hindu astronomy.

"Three Indian astronomical texts are cited by the first generation of Arab scientists: Aryabhatiya [*Āryabhaṭīya*], written by Aryabhata [Aryabhaṭa I] in 499 [CE] and referred to by Arab authors under the title *al-arjabhar*; *Khandakhadyaka* by Brahmgupta (598–668 CE), known in Arabic under the title *Zīj al-arkand*; and *Mahasidhanta* [*Mahāsiddhānta*], written toward the end of the seventh or at the beginning of the eighth century, which passed into Arabic under the title *Zīj al-Sindhind*," writes Régis Morelon.[86] A multitude of *Zīj*s were written in India and Afghanistan first, and in Persia and Baghdad later. A typical *Zīj* covered information on trigonometry; spherical astronomy; solar, lunar, and planetary mean motions; solar, lunar and planetary latitudes, parallax, solar and lunar eclipses, and geographical coordinates of various locations, particularly to locate *qibla*.[87] "The Arabic text is lost and the work has been transmitted through a Latin translation made in the twelfth century by Adelard of Bath from a revision made in Andalusia by al-Majrīṭī (d. 1007 CE)."[88]

Al-Khwārizmī even used metamorphosed Sanskrit terms in his astronomical calculations. For example, for the rules when finding the sizes of the Sun, the Moon and the Earth's shadow, al-Khwārizmī used the term *elbuht*, that comes from the Sanskrit word *bhukti* where the shadows on the Earth from the Sun and the Moon was observed at the same time daily, indicating the mathematical processes were essentially taken from Hindu astronomy.[89]

Ṣā'id al-Andalusī (1029–1070 CE) writes that "a person originally from Hind came to Caliph al-Manṣūr in A.H. 156 [773 CE] and presented him with the arithmetic known as *Sindhind* for calculating the motion of stars. It contains *ta'ādyal* [equations] that give the positions of stars with an accuracy of one-fourth of a degree. It also contains examples of celestial activities such as the eclipses and the rise of the zodiac and other information. . . . Al-Manṣūr ordered that the book be translated into Arabic so that it could be used by Arab astronomers as the foundation for understanding celestial motions. Muhammad ibn Ibrahim al-Fazārī accepted the charge and extracted from the book that astronomers called *al-Sindhind*. . . . This book was used by astronomers until the time of Caliph al-Ma'mūn, when it was abbreviated for him by Abu Ja'far Muḥammad ibn Mūsā al-Khwārizmī, who extracted from it his famous tables, which were commonly used in the Islamic world."[90]

[86] Régis Morelon, General Survey of Arabic Astronomy, in the book by Roshdi Rashed, 1996, vol. 1, p. 8.

[87] King, in the book by Selin, 1997, p. 128; Mercier in Selin, 1997, p. 1057. Zij is a common term used in Arabic for tables and *qibla* is the direction of Kaaba (Mecca) from your location.

[88] Régis Morelon, Eastern Arabic Astronomy Between the Eighth and the Eleventh Centuries in the book by Roshdi Rashed, 1996, vol. 1, p. 21; see also Toomer, G. J., 1973, *Dictionary of Scientific Biographies*, vol. 7, p. 360.

[89] Neugebauer, 1962, p. 57; Goldstein, 1996.

[90] Salem and Kumar, 1991, p. 46–47. This tells us that Europeans were aware that al-Khwārizmī's work was taken from Hindu astronomers.

Caliph al-Manṣūr was a ruler of Baghdad and his Abbasid dynasty was known for its respect of knowledge. He established *Bayt al-Hikma* (House of Wisdom) which became a model for other empires in Arabia and Europe. The House of Wisdom was a court or school where scholars worked in history, jurisprudence, astronomy, mathematics, and medicine, etc. These scholars were supported by Caliph al-Manṣūr and, in return, they helped the Caliph in his personal and kingdom affairs. It was a practice of the rulers of Abbasid dynasty to patronize scholars from foreign lands. With time, Baghdad became a center of learning. Scholars from the nearby regions visited Baghdad to acquire knowledge.

Abū Ma'sher, Jafar ibn Moḥammad ibn Amar al-Balkhī (787–886 CE), a Persian astronomer who mostly lived in Baghdad, also mentioned Kanaka's role in Baghdad. He labeled Kanaka as the foremost astronomer among all the Indians of all times.[91] We do not know much about Kanaka. Most of the information about him has come from the manuscripts written later in Arabia, the Mediterranean region and Europe. Apparently, Kanaka made his impact outside India.

In Persia, under the Sasanids (226–651 CE), observational astronomy was practiced under the influence of Indian and Greek astronomy. We know from al-Hashīmī (fl. ninth century) that Shāh Anūshirwān compared the work of Ārybahaṭa I's *Arkand* with Ptolemy's *Almagest*. He found Āryabhaṭa I's work better than Ptolemy's. Thus, the king asked his astronomers to compile a *Zīj* on Āryabhaṭa I's system. This is how the "Royal tables" (*Zīj al-Shāh*) were compiled.[92] Al-Manṣur, while deciding the auspicious time for the foundation of the capital Baghdad, asked his astronomers to use a Pahlavi version of *Zīj al-Shāh* to calculate this time.[93]

Like the Greenwich observatory in England has become a standard location to define the time and longitude of various locations in the world, al-Khwārizmī used Arin (Ujjain), the Greenwich of the ancient and medieval worlds, as the central place of the Earth.[94] This is an important piece of information. Almost any point on the Earth can be chosen as a standard for this. Al-Khwārizmī could have chosen Baghdad as the prime meridian, his place of residence. However perhaps due to the prevalent practice in Arabia and al-Khwārizmī's dependence on the Hindu astronomical tables, he preferred to choose Ujjain. It is the city that was also chosen by Āryabhaṭa I and Brahmgupta, from whom al-Khwārizmī derived his work, as zero meridian. al-Khwārizmī defined one sidereal year equal to 365.15302230 days. This is exactly the same value used by Brahmgupta.[95] In Al-Khwārizmī's book, the "era of flood" was the era of *Kaliyuga* (February 17, 3102 BCE). Al-Khwārizmī's *elwazat*, a procedure to calculate the mean positions of the planets, was similar to the *ahargaṇa* method of Hindu astronomy.[96]

[91] Pingree, 1968, p. 16.

[92] van der Waerden, The Heliocentric System in Greek, Persian, and Hindu Astronomy, in the book by King and Saliba, 1987; Régis Morelon, General Survey of Arabic Astronomy, in the book by Roshdi Rashed, 1996, vol. 1, p. 8. On a different note, under Anūshirwān's reign, chess was introduced from India, and the famous book, *Kalilah and Dimnah* was translated.

[93] David King, in the book by Selin, 1997, p. 126; F. Jamil Ragep, in the book by Selin, 1997, p. 395.

[94] Neugebauer, 1962, p. 10, 11.

[95] Neugebauer, 1962, p. 131.

[96] Sen, 1970.

4.4.2 CHINA

The influence of the Hindu thought on China can be judged by the fact that, between 67 CE and 518 CE, in less than five centuries, some 1,432 works in 3,741 fasciculi were translated from Sanskrit to Chinese and were cataloged in 20 disciplines.[97] By 433 BCE, the Chinese astronomers recorded a system of 28 hsiu (lunar mansions or Moon stations) marked by a prominent star or constellation. The Moon traveled past and lodged in each of these mansions. This system probably originated from the Hindu system of 28 *nakṣatra* (star or constellation).[98] The *Vedas* provide a complete list of these *nakṣatra*. One difference between these two systems is that whereas the Hindus named their *nakṣatra* after their gods, the Chinese honored their emperors, queens, princes, and even bureaucrats by assigning their names to stars.

The Navagraha-siddhānta (*Nine Planet Rule*), a popular Indian astronomy book, was translated into Chinese as *Kaiyuan Zhanjing* in 718 CE by Indian astronomer Gautama Siddhārtha (Qutan Xida, his Chinese name). This book is still preserved in a collection from the Tang period. It contains the Indian calendar, known as *navagraha* (nine houses; used for the five planets, the Sun, the Moon, and the two nodal points, Rāhu and Ketu). This calendar was known as *Jiuzhi li* (Nine Upholder Calendar) in China.[99] This calendar is based on Varāhamihira's *Pañca Siddhānta* that was written during the sixth century. It contains the astronomical tables along with the methods to calculate an eclipse. It has a table of sines and the Hindu numeral zero in the form of a dot (*bindu*).[100] The Chinese, in following the Hindu tradition, also used nine planets in their astronomical work that included the Sun, the Moon, Mercury, Venus, Saturn, Jupiter, Mars, and the ascending and descending nodes known as Rāhu and Ketu in the Hindu work.[101] In May 1977, a tomb of an Indian astronomer from the Gautama family was excavated at Chang-an county in modern Shaanxi province. This tomb had manuscripts and inscriptions that has provided valuable information about Gautama clan. As a result of this excavation, we now know that Gautama Siddhārtha was not the only famous person in Gautama clan. Gautama Zhuan also played an important role in the Bureau of Astronomy and at the Tang court. He got married to a Chinese woman and the following generations were assimilated into the Chinese culture.[102]

During the Tang Dynasty (618–907 CE), three schools of Indian astronomical systems were based in China to guide the emperors. These schools are: Siddhārtha school, Kumāra school, and Kaśyapa school.[103] At least two experts from the Siddhārtha school served as Director of the astronomical bureau during Tang dynasty. It is important to mention that these directors were crucial for emperors in their day-to-day activities as they had to find auspicious

[97] Gupta, 1981; Mukherjee, 1928, p. 32, taken from Gupta, 1981.
[98] Sarma, 2000.
[99] Yoke, 1985, p. 162
[100] Yoke, 1985, p. 162.
[101] Bagchi, 1950, p. 169 and Yoke in Selin, 1997, p. 78.
[102] Sen, Tansen, 1995.
[103] Yoke in Selin, 1997, p. 110.

times for rituals and government actions, consult the director on astrological matters, and even take advice on dealing with people. This is perhaps the most powerful position, after the emperor himself. This position was occupied by Hindus for several generations during the Tang dynasty. This fact tells a lot about the status of Hindu astronomy and Hindus in China.

Amoghavajra (Chinese name, Bukong), a brahmin from India arrived in China in 719 at the age of 15 with his uncle. In 741, he went back to India and again returned to China in 746. He was given the title of Zbizang by the Tang Emperor Xuanzong (713–756). In 759 CE, Amoghavajra wrote *Xiu yao jing* (Lunar mansion and planet sutra). A commentary to this work by Yang Jingfeng mentioned the three Indian schools of astronomy prevalent in China and occupants of powerful Bureau of Astronomy: "Those who wish to know the positions of the five planets adopt Indian calendrical methods. One can thus predict what hsiu (a planet will be traversing). So we have the three clans of Indian calendar experts, Chia yeh (Jia ye Kasyapa), Chhu than (Qutan Gautama) and Chu mo lo (Jiu mo lo Kumara), all of them hold office at the Bureau of Astronomy. But now most use is made of the calendrical methods of Master Chhu than chuan (Qutan zhuan) together with his Great Art."[104] Another person from this clan, Gautama Rahula, was the director of astronomy between 627 to 649, and compiled two calendars viz. *Jingweili* and *Guangzaili*. As the family settled down in China, marry the Chinese, they are not labeled as Indian in Chinese records.

Rāhu (*Chiao chhu*) and Ketu (*chiao chung*) were frequently mentioned in the Chinese texts written during and after the Tang Dynasty (618–907 CE).[105] The book, *Qi Yao Rang Zai Jue* (*Formulae for Warding off Calamities According to the Seven Luminaries*), was compiled by an Indian Buddhist monk, Jin Ju Zha, in China. This book has detailed ephemerides of Rāhu and Ketu. According to this book: "The luminary E Luo Shi Rāhu is also known by the following names: The Yellow Standard (*Huang Fan*), The Head of the God of Eclipse (*Shi Shen Tou*), Superposition (*Fu*), and The Head of the Sun (*Tai Yang Shou*). It always moves invisibly and is never seen when it meets the Sun or the Moon, an eclipse occurs; if the meeting is at a new moon or a full moon, then an eclipse necessarily occurs; when it is opposite to the Sun or the Moon, there will also be an eclipse."[106] In summary, the text supports the following points:

1. Rāhu and Ketu are invisible luminaries.

2. The motion of Rāhu and Ketu have a bearing on the occurrence of eclipses.

3. The theory depicted in the tales of Rāhu and Ketu is different than the ancient theory of eclipses in China.

4. Rāhu and Ketu execute a uniform motion against the background of fixed stars and its speed does not vary.

[104]Deshpande, 2015; Sen, 1995. A similar account is also provided by Needham and Ling, 1959, vol. 3, p. 202 and Yoke, in Selin, 1997, p. 78.

[105]Needham and Ling, 1959, vol. 3, p. 416.

[106]Wei-xing, 1995.

4.4.3 EUROPE

As mentioned earlier, during the eleventh century, the European scholars in Spain knew that al-Khwārizmī's work in the Middle East was an extension of Hindu astronomy. Ṣā'id al-Andalusī, a prominent scholar from Spain, provided ample information about the work of Āryabhaṭa I, Brahmgupta, etc. He believed that the Hindus' work on astronomy formed the basis for Arab astronomy,[107] as mentioned in the previous subsection. Ṣā'id al-Andalusī was not the only person stating this fact. Rabbi Abraham Ibn Ezra (1096–1167 CE) provided a similar story of this transfer of knowledge from India to Arabia.[108] Ibn Izra was based in Spain and wrote about Hindu mathematician and astronomer, Kanaka, who shared his knowledge and allowed the Arabs to know about Hindu astronomy: "The scholar, whose name was Kanaka, was brought to the king, and he taught the Arabs the basis of numbers, i.e., the nine numerals. Then, from this scholar with the Jew as Arabic-Indian interpreter, a scholar named Jacob b. Sharah translated a book containing the tables of seven planets [five planets, the Sun, and the Moon]. . ."[109]

Abū Ishaq Ibrāhim ibn Yahya al-Naqqash, better known as al-Zarqalī (ca.1029–1087 CE), a Spanish astronomer who worked under Ṣā'id al-Andalusī, compiled the famous Toledan Tables that are based on *the Sindhind* system.[110] "One of the first Latin authors to use tables of Arabic origin was Raymond of Marseilles. In 1141 [CE], he composed a work on the motions of the planets, consisting of tables preceded by canons and an introduction in which he claims to draw on al-Zarqāllu [Zarqalī].[111]

Prior to al-Zarqalī, around 960 CE, 'Arīb bin Sa'īd and Mozarab bishop Rabī b. Zayd compiled the Calendar of Códoba for al-Ḥakam II after his accession to the caliphate. The calendar provided the dates when the Sun enters in different zodiacs. These dates are provided according to the Sindhind and used the mathematics of Brahmgupta.[112]

"The Toledan tables are a composite collection, including the parts taken from the tables of al-Zarqāllu [Zarqalī] alongside extracts from al-Khwārizmī (notably the planetary latitudes), elements from Al-Battānī (in particular, the tables of planetary equations), and yet other parts derived from the *Almagest* or the *Handy Tables* of Ptolemy."[113] The *Toledan tables* were translated by Gerard of Cremona[114] and played an important role in the growth of European astronomy.

Al-Zarqalī is not the only person in Europe who studied and wrote astronomy books on the Hindu system of *Sindhind.* Al-Majrīṭī (d. 1007 CE) also modified the *Sindhind* of al-

[107] Salem and Kumar, 1991

[108] Goldstein, 1967, p. 1478.

[109] Goldstein, 1996.

[110] Sarton, 1927, vol. I, p. 759.

[111] Henri Hugonnard-Roche, Influence of Arabic Astronomy in the Medieval West, in the book by Roshdi Rashed,1996, vol. 1, p. 287; also see, Toomer, 1968.

[112] Juan Vernet and Julio Samsó, Development of Arabic Science in Andalusia, in the book by Roshdi Rashed,1996, vol. 1, p. 250–251.

[113] Henri Hugonnard-Roche, Influence of Arabic Astronomy in the Medieval West, in the book by Roshdi Rashed, 1996, vol. 1, p. 289.

[114] Henri Hugonnard-Roche, Influence of Arabic Astronomy in the Medieval West, in the book by Roshdi Rashed, 1996, vol. 1, p. 292.

Khwārizmī. A student of al-Majrīṭī, Ibn al-Samḥ (979–1035 CE), composed a *zij* that was based on the *Sindhind* system.[115] Another student of al-Majrīṭī, Ibn al-Saffār, also authored a brief astronomical table that was based on the *Sindhind* system."[116] The third scholar from Spain, Ibn al-Ādāmi also compiled the astronomical tables known as *Kitab Naẓm al-'Iqd* (*Book on the Organization of the Necklace*) that was completed after his death by his student al-Qāssim bin Moḥammad ibn Hashim al Madā'inī, better known as al-'Alawī, in 950 CE. "This book contains all that was known about astronomy and the calculation of the motions of the stars according to the system of the *Sindhind*, including certain aspects of the trepidant motions of the celestial bodies, which were never mentioned before," writes Ṣā'id al-Andalusī.[117]

It is the Latin version of al-Khwārizmī's text that was used by Adelard of Bath, a British scientist and Arabist of the early twelfth century, to teach Henry Plantagenet, the future King Henry II of England.[118] Thus, the work of the Brahmins on the banks of Ganges became the subject matter of learning to the Royalty in England.

Alfonso X, King of Castile[119] (1252–1284 CE) gathered a group of Muslims, Jewish, and Christian scholars in the tradition of *Bayt al Hikma* or *House of Wisdom* of Baghdad. These scholars translated the earlier works in Arabic that were based on Hindu astronomy and compiled them into a single book in the Castilian language, called the Alfonsine Tables. This book reached Paris in the early 14th century and was translated into Latin. Soon the book spread throughout Europe and used as computing tool by European astronomers.[120] Thus, medieval astronomy in Europe was a derived and improved mixture of Hindu, Persian, and Arabic contributions. It is established that a bound volume of the Alfonsine Tables was owned by Copernicus and he did copy planetary latitudes from these tables.[121] The extent to which Copernicus was in debt of the Hindu science is a matter of interpretation and debate. However, there is no dispute that he did use Hindu numerals and mathematics in his calculations. His dependence on Hindu astronomy is an open question which only the future scholarships will ascertain.

Johannes Kepler gave his table of parallax of the Moon which is essentially identical with the theory of parallax given in *Khandakhadyaka* of Brahmgupta.[122] The theory of Brahmgupta was proposed about one millennium before Kepler. A possible connection could be al-Khwārizmī's book on Hindu astronomy, *Zīj al-Sindhind*. This book quite popular in Europe when Kepler was learning astronomy in his youth.

[115]Salem and Kumar, 1991, p. 64.

[116]Salem and Kumar, 1991, p. 65.

[117]Salem and Kumar, 1991, p. 53.

[118]Burnett, 1997, p. 31.

[119]Castile is presently a part of Spain. As a side note, soaps made from olive oil and sodium hydroxide or any hard soap from fat or oil are called Castile soaps. It is the legacy of this region.

[120]Chabás, 2002; King and Saliba in the book by Gingerich, 1993.

[121]Swerdlow and Neugebauer, 1984, p. 4 and Chabás, 2002.

[122]Neugebauer, 1962, p. 124. This is the conclusion of Neugebauer (1899–1990), a noted historian of science, who was trained in Gröningen, Germany, and taught at Brown University, USA.

CHAPTER 5

Physics

Physics deals with matter and energy and their interactions. Measurements are central to the growth of physics and length (space), time, and mass, are the three most important physical quantities, called the fundamental quantities. Most other physical quantities are generally expressed in their terms of mass, length, and time. For example, speed is measured in miles per hour (or kilometer per hour) and involves a measurement of space (distance) and time. This means that a car moving with 65 miles/hour moves 65 miles (one measurement) in one hour (second measurement). Similarly, force is measured in terms of mass, length, and inverse-square of time. Therefore, for convenience purposes, most civilization defined standards for these fundamental quantities. The ancient Hindus also methodically and carefully defined these standards.

5.1 SPACE (*ĀKĀŚA*)

Space is a three dimensional matrix into which all objects are situated and move without producing any interaction between the object and the space. In the classical sense, space allows a physical ordering of objects without any reference to time. Objects appear to be near or distant due to this physical ordering. In the Newtonian (classical) world, where objects move with speeds much smaller in comparison to the speed of light, space and time are independent of each other, and are considered as separate fundamental quantities. In the relativistic world, where objects move with a velocity that is comparable to the velocity of light (3×10^8 m/s), space and time do not have their independent status; they are integrated and form a new space-time (spatio-temporal) reality. In the present context, only the classical picture of space and time as two independent realities are considered.

Ākāśa is one of the terms used to describe "space" in the Sanskrit language. According to the *Chāndogya-Upaniṣad*, space was the first entity in the creation of the universe. To a question, "To what does the world go back?," the *Chāndogya-Upaniṣad* answers: "To space, all things are created from space and they dissolve into space. Space alone is greater than any manifestation; space is the final goal."[1] The *Chāndogya-Upaniṣad* identified space as an entity that contains the Sun, the Moon, and other material objects: "In Space are both the Sun and the Moon, lightening, the stars, and fire. Through space one calls out; through Space one hears; through Space one answers. In Space one enjoys himself. In space one is born. Reverence Space."[2] Thus, even the heavenly objects were a part of something larger and more basic than them; this was space

[1] *Chāndogya-Upaniṣad*, 1: 9: 1.
[2] *Chāndogya-Upaniṣad*, 7: 13: 1–2.

that encompassed humans, sound, the Sun, the Moon, and the stars. It is space that provides individuality to objects as they are recognized based on the space they occupy: "Space is the accomplisher of name and form (individuality)."[3]

The *Vaiśeṣika-Sūtra* of Kaṇāda explained the attribute of space: "That which gives rise to such cognition and usage as, this is remote from this, is the mark of space."[4] Thus, distance is an attribute of space. Kaṇāda considered the various directions (east, west, north, south) as the attributes of space.[5]

A standard of length is imperative for any measurement of space. The *Mārkaṇdaya-purāṇa* provides the following standards of length: "A minute atom, a *para-sukṣma*, the mote in a sunbeam, the dust of the earth, and the point of a hair, and a young louse, and a louse, and the body of a barley-corn; men say each of those things is eight times the size of the preceding thing. Eight barley-corns equal an "aṅgula" (finger-breadth), six finger-breaths are a "pada" (step), and twice that is known as bitasti; and two spans make a cubit measured with the fingers closed in at the root of the thumb; four cubits make a bow, a "daṇḍa" (stick), and equal to two "nadikayaga;" two thousand bows make a "gavyūti;" and four times that are declared by the wise to be a "yojna;" this is the utmost measure for the purpose of calculation."[6]

In Sanskrit, "pāda," literally meaning a human step. If we consider "pāda" to be equal to one foot and calculate the size of an atom using *Mārkaṇdaya-purāṇa*, its value comes out to be 2.9×10^{-9} meter which is strikingly closer to the actual value (of the order of 10^{-10} m).

Kauṭilaya, the teacher and Prime Minister of Chandragupta Maurya (Candragupta, reigned, 322–298 BCE) defines several standards of length in his book, *Arthaśāstra*[7] that are compiled in Table 5.1.

5.2 TIME

It is easy to measure time. Most people carry a watch and some of these watches can be bought for just a few dollars. To define the nature of time is comparatively a lot difficult task. Despite consistent efforts for at least two millenniums, it is still an open question. In other words, it is a lot easier to set a standard for time; it is a lot difficult to set an abstract definition of time that can exemplify its nature.[8]

Time and action (event) are related with each other. Events are the basis of time. Time deals with the order and the duration of events. Since events are discrete in nature, this makes time discrete in nature. If there is a sequence of events at a definite time interval, careful measurements of such a body constitutes a clock. In simple words, time divides two events from one another. What happens when events move faster and the time between the two events becomes

[3] *Chāndogya-Upaniṣad*, 7: 14: 1.
[4] *Vaiśeṣika-Sūtra*, 2: 2: 10.
[5] *Vaiśeṣika-Sūtra*, 2: 2: 11–13. For this book, I have used a Hindi edition by Pundit Shriram Sharma Āčarya.
[6] *Mārkaṇdeya-purāṇa*, 49: 37.
[7] *Arthaśāstra*, Chapters 106–107; Book 2, Chapter 20; Shamasastry, 1960, p. 117.
[8] For the history of time and clocks, read Balslev, 1983; Panikkar, 1984; and Prasad, 1992.

Table 5.1: A Standard of Length from Kauṭilaya's *Arthaśāstra*

8 paramāṇavah, atoms	1 particle thrown off by the wheel of a chariot
8 particles	1 likṣā (egg of louse or young louse)
8 likṣā	1 yūka (louse of medium size)
8 yūka	1 yava (barley of medium size)
8 yava	1 aṅgula (finger breadth)
4 *aṅgula*	1 dhanurgṛaha
8 aṅgula	1 dhanurmuṣaṭi
12 aṅgula	1 vitasti
14 aṅgula	1 śamya or pāda
2 vitasti	1 aratni or prajāpatya hasta (arm-length)
42 aṅgula	1 kiṣku (forearm)
54 aṅgula	1 hasta used in measuring timber forests.
84 aṅgula	1 vyama
4 aratini	1 rajju (rope)
10 daṇḍa	1 rajju
2 rajju	1 parideśa
3 rajju	1 nivartana
66.5 nivartana	1 goruta
4 goruta	1 yojna.

smaller and smaller? Do we have continuous events and, therefore, continuous time? Philosophers have considered the question as to whether time is continuous or discrete for ages, without coming to a definite conclusion.

Any temporal existence is evolved with time and vice-versa. It is time that is the origin of the first cause, as the *Atharvaveda* tells us: "Time created living things and first of all, Prajāpati."[9] In Hindu scriptures, "Prajāpati" is the first God created who, in turn, created the present universe. In this view, it is time that unfolds the spatio-temporal universe and divides the past from the present and the future. "In time, texts are produced—what is, and what is yet to come."[10] "Time is the supreme power in the universe."[11]

The *Maitreyī-Upaniṣad* relates time to the *mūrta* (existent) world. "There are two forms of Brahmā: Time and Timeless. That which is before the Sun is Timeless and that which begins

[9] *Atharvaveda*, 19: 53: 10.
[10] *Atharvaveda*, 19: 54: 3.
[11] *Atharvaveda*, 19: 54: 6.

with the Sun is Time."[12] This quotation states that the Sun was created with the creation of universe and that time existed only after the creation. Prior to the creation, time did not exist, in the absence of any event. "[A]nything before the Sun" defines the period before the physical manifestation of the world, the situation before the Big Bang, when matter was in noumenal form, and no events were possible. Therefore, time could not have existed in the absence of events. Only after the creation, matter became accessible to experience.

The *Vaiśeṣika-Sūtra* considered time as an entity that exists only in the manifested world (non-eternal), like in the *Maitreyī-Upaniṣad* mentioned above. "The name Time is applicable to a cause, in as much as it does not exist in eternal substances and exists in noneternal substances."[13] Therefore, time can only be noticed in a dynamic world (temporal world) where events are happening and serve as distinguishing factors. In a void, after this temporal world dissolves into "darkness" or "non- existence," time cannot exist.[14]

The *Viṣṇudharmotra-Purāṇa* explains the subtlety of time and its quantized nature: "if one pierces 1,000 lotus petals (put on top of each other) with a needle, the foolish man thinks that they are pierced simultaneously, but in reality they were pierced one after the other and the subtle difference between the instants in which the successive petals have been pierced represents the subtlety of time."[15]

Āryabhaṭa I explains the nature of time and defines a method to measure it: "Time, which has no beginning and no end, is measured by (the movements of) the planets and the asterisms on the sphere."[16] To measure time, the apparent motion of the Sun defined the events; sunrise and sunset were two easily observed events. A scale based on the average-solar day was established by the ancient Hindus. The duration of an average day was divided into several segments that became a standard of time.

To measure the magnitude of time, a simple method is defined in the *Srimad-Bhāgvatam*.[17] (Table 5.2.) The duration of "nāḍikā" is measured as follows: "Take a copper vessel measuring six "pala;" make a hole into the copper vessel by a pin made of gold of which the length shall be four fingers and measure four "maśa" (unit of mass). Put the vessel on water. The time taken to make the vessel filled with the water and sink constitutes one nāḍikā." If we consider the average length of a day to be 12 hours then the smallest unit of time, "trasareṇu," comes out to be about 1.7×10^{-4} seconds.[18]

Within the accuracy of a solar clock, the method provided in the *Srimad-Bhāgvatam* for the measurement of time is fairly good. All factors that can influence a measurement are provided—the size of the copper vessel, the size of the hole is defined from the length of the gold pin and its weight that defines the diameter of the pin, and the type of liquid.

[12] *Maitreyī-Upaniṣad*, 6: 15.
[13] *Vaiśeṣika-Sūtra*, 2: 2: 9.
[14] see *Ṛgveda*, 10: 129: 1–4; see Chapter 4.
[15] *Viṣṇudharmotra-Purāṇa*, 1: 72: 4–6.
[16] *Āryabhaṭīya*, *Kalākriya*, 11.
[17] *Srimad-Bhāgvatam*, 3: 11: 6–11.
[18] *Srimad-Bhāgvatam*, 3: 11: 6–11.

Table 5.2: A Standard of Time from *Srimad-Bhāgvatam*

3 trasareṇu	1 truti
100 truti	1 bedha
3 bedha	1 lāva
3 lava	1 nimeṣa
3 nimeṣa	1 kṣaṇa
5 kṣaṇa	1 kāṣṭha
15 kāṣṭha	1 laghu
15 laghu	1 nāḍikā or daṇḍa
2 nāḍikā	1 muhūrta
6 or 7 daṇḍa	1 prahara (one-fourth of a day or night)

Table 5.3 defines a standard of time provided in Kauṭilaya's *Arthaśāstra*.[19] One *nālika* is defined as the time during which one *adhaka* (a measurement of mass) of water passes out of a pot through an aperture of the same diameter as that of a wire of 4 *aṅgula* and made of 4 *maśa* (measurement of mass) of gold. The length and mass of gold defines the diameter of the wire. Here, truṭi comes out to be 0.06 s, if we assume 1 day to be 12 hours.

Table 5.3: A Standard of Time from Kauṭilaya's *Arthaśāstra*

2 truṭi	1 lava
2 lava	1 nimeṣa
5 nimeṣa	1 kāṣṭha
30 kāṣṭhā	1 kalā
40 kalā	1 nālika
2 nālika	1 muhūrta
15 muhūrta	1 day or night

Al-Bīrūnī explained that the smallest unit of time used in India was *aṇu* that is equal to $(24 \times 60 \times 60)/(88,473,600) = 9.766 \times 10^{-4}$ seconds.[20] He did not see a use of such a small unit for time, just like he did not like the Hindus to define large numbers. "The Hindus are foolishly painstaking in inventing the most minute segment of time, but their efforts have not resulted in a universally adopted and uniform system."[21] With nano-second (10^{-9}s) or femto-

[19] *Arthaśāstra*, 107; Shamasastry, 1960, p. 119.
[20] Sachau, 1964, vol. 1, p. 337.
[21] Sachau, 1964, vol. 1, p. 334.

second (10^{-15}s) prevalent in scientific works today, it simply shows that the Hindus were simply ahead of their time.

5.3 MATTER AND MASS

In order to define and distinguish matter, Kaṇāda defines matter (*padārtha*) into six categories.[22]

1. Substance (*dravya*)
2. Quality (*guṇa*)
3. Action (*karma*)
4. Generality (*sāmānya*)
5. Individuality (*viśeṣa*)
6. Inherence (*samavāya*)

Matters of different materials have different attributes. For example, identical sizes of iron and cotton would have different weights. Apples can have differences in their colors and yet be called apples. It is the assembly of attributes that defines matter. Table 5.4 defines the attributes of matter as defined by Kaṇāda.

Table 5.4: Attributes of matter from *Vaiśeṣika-Sūtra* (1: 1: 6)

1.	color (*rūpa*)	10.	priority (*paratva*)
2.	taste (*rasa*)	11.	posteriority (*aparatva*)
3.	smell (*gandha*)	12.	intellect (*buddhi*)
4.	touch (*sparśa*)	13.	pleasure (*sukha*)
5.	number (*saṅkhyā*)	14.	pain (*duḥkha*)
6.	measure (*parimāṇa*)	15.	desire (*icchā*)
7.	individuality (*pṛthaktā*)	16.	aversion (*dveṣa*)
8.	conjunction (*saṁyoga*)	17.	volition (*prayatnaḥ*)
9.	disjunction (*vibhāga*)		

To explain what he meant by different attributes of matters, Kaṇāda provides examples: Earth has the attributes of color, taste, and touch;[23] water has the attributes of color, taste, touch, fluidity, and viscidity;[24] fire with color, and touch;[25] air has the attribute of touch;[26] while space

[22] *Vaiśeṣika-Sūtra*, 1: 1: 4.
[23] *Vaiśeṣika-Sūtra*, 2: 1: 1.
[24] *Vaiśeṣika-Sūtra*, 2: 1: 2.
[25] *Vaiśeṣika-Sūtra*, 2; 1: 3.
[26] *Vaiśeṣika-Sūtra*, 2; 1: 4.

has no such properties.[27] While analyzing air, Kaṇāda concludes that "air is a substance since it has action and attributes."[28] "Air is matter, but its non-perception, in spite of being substance, is due to the non-existence of color in it."[29] With the non-perception of air, Kaṇāda defines reality beyond appearance. Not all that exists in the world is visible to human eyes. In his attempt to clarify a difference between space and air, Kaṇāda used the attribute of touch which is absent in space: "Air has the property of touch while space does not have such a property."[30] To support that air is matter, Kaṇāda argued that a breeze can easily move the leaves of grass.[31] Thus, air, though invisible, can exert force and move things.

Kauṭilaya's *Arthaśāstra* provided standards of mass and suggested that the standards should be made of iron or stones, available in Magdha and Mekala, both parts of Central India. A replacement could be of objects that do not change their physical conditions. These standards should not contract or expand when wet, and a change of temperature should not affect them.[32]

For a measurement of mass, Kauṭilaya defined balances with different lever arms and scale-pans for a range of masses:[33] With the fulcrum in the middle, two identical pans hang on both sides of the fulcrum at equal distances. Standard appropriate masses are placed on one side and the unknown mass on the other so that the beam is balanced when the device is suspended by the fulcrum. Kauṭilaya provides descriptions of sixteen types of such balances. He suggested different balances for different accuracy: "public-balance" (*vyāvahārikā*), "servant-balance" (*bhājini*) and "harem-balance" (*antaḥpura-bhājini*).[34] Superintendents were assigned to stamp mass-standards for public use to avoid cheating, and money was charged for such services.[35] Traders were not allowed to use their own standards, and were fined if found guilty of doing so.[36]

Much before Kauṭilaya, a beam balance with two identical pans of metallic copper or bronze hanging from each end of the beam and a fulcrum in the middle have been found in Harappa and Mohenjodaro archaeological excavations. The beams are thicker in the middle around fulcrum and tapered at the end where pans are suspended. Most pans have three symmetrical holes to suspend using strings. The diameter of these pans range from 5.50 to 8.25 cm. From Lothal, Gujrat, even teracotta pans have been excavated. Some micro-balances with standard weights ranging from 0.89 to 1.2 g were also excavated.[37]

[27] *Vaiśeṣika-Sūtra*, 2; 1: 5.
[28] *Vaiśeṣika-Sūtra*, 2: 1: 12.
[29] *Vaiśeṣika-Sūtra*, 4: 1: 7.
[30] *Vaiśeṣika-Sūtra*, 2: 1: 4, 5.
[31] *Vaiśeṣika-Sūtra*, 5: 1: 14.
[32] *Arthaśāstra*, 103; Book 2, Chapter 19; Shamashastri, 1960, p. 113.
[33] *Arthaśāstra*, 103; Book 2, Chapter 19; Shamasastri, 1960, p. 114.
[34] *Arthaśāstra*, 104; Book 2, Chapter 19; Shamasastri, 1960, p. 114.
[35] *Arthaśāstra*, 105; Book 2, Chapter 19; Shamasastri, 1960, p. 116.
[36] *Arthaśāstra*, 105; Book 2, Chapter 19; Shamasastri, 1960, p. 116.
[37] Sharma and Bhardwaj, 1989.

5.3.1 CONSERVATION OF MATTER

Conservation of matter is a fundamental law in physics that was introduced by John Dalton (1766–1844 CE) in 1803. Einstein's mass-energy relation added energy to the conservation of mass since mass can be transformed into energy and vice-versa. Mass and energy are two different manifestations of the same reality. Therefore, the law of conservation of matter is now transformed into a broader law of conservation of energy to incorporate mass-energy transformations.

Conservation of matter, in the ancient Hindu literature, is an extension of the theory of reincarnation. The ancient Hindus noted that life is a continual process and birth and death are just two different stages of life. In an extension of birth and death to the inorganic world, they believed that matter, like the soul, cannot be destroyed or created; only a transformation from one form to another is possible. As discussed in Chapter 4, the possibility of creation from "nothing" is rejected by the ancient Hindus.

The *Vaiśeṣika-Sūtra* tells us that a substance is produced from other substances.[38] Mixing or separation of different substances creates new substances and it is impossible to create matter from "nothing."[39] In his attempt to explain the properties of air, Kaṇāda suggests the law of conservation of matter that is strikingly similar to the definition accepted by the scientists about hundred years ago. "Matter is conserved. Since air and fire are made up of atoms, these are also conserved."[40] Kanada also makes an explicit statement that "molecules and materials are eternal," implying that they are conserved.[41] The Sanskrit term *nitya* is used to define that atoms and matter are conserved. The most common translation of *nitya* is "eternal," defining something that will exist forever. This essentially means that atoms are eternal or conserved.

5.4 ATOM (*PARAMĀṆU*)

Karl Marx (1818–1883), the father of communism, wrote his doctoral thesis on the differences of the philosophies of Democritus and Epicurus in 1841. While studying the life of Democritus, Karl Marx suggested that Democritus came in contact with Indian gymnosophists and the concept of atom was already present in India during that period. Karl Marx was not the first person to indicate this. The connection of the Greek philosophers with India has been accepted by some scholars through out in antiquity and the medieval world. This is also the view of Diogenes Lucretius (First Century BCE; IX: 35) whom Marx studied extensively for his work. Marx writes, "Demetrius in the *Homonymois* [*Men of the Same Name*] and Antisthenes in the *Diadochais* [*Successions of Philosophers*] report that he [Democritus] traveled to Egypt to the priests in order to

[38] *Vaiśeṣika-Sūtra*, 1: 1: 23 and 27.
[39] *Vaiśeṣika-Sūtra*, 1: 1: 25 and 28.
[40] *Vaiśeṣika-Sūtra*, 7: 1: 4; 7: 1: 19–20.
[41] *Vaiśeṣika-Sūtra*, 7: 1: 8.

learn geometry, and to the Chaldeans in Persia, and that he reached the Red Sea. Some maintain that he also met the gymnosophists in India . . ."[42]

Kaṇāda's atomic theory was formulated sometime between the sixth century BCE and tenth century BCE.[43] Kaṇāda was from Pabhosa, Allahabad, and was perhaps the first atomist in India.[44] His theory was quite popular in ancient and medieval India. The *Vāyu-Purāṇa*, *Padma-Purāṇa*, *Mahābhārata*, and *Srimad-Bhāgvatam* list Kaṇāda's work and bear infallible testimony to the antiquity and popularity of the *Vaiśeṣika-Sūtra*.

In Kaṇāda's theory, the combination of molecules (*aṇu*) is possible and the non-perception of atoms disappears when they amass together to become bigger.[45] This indirectly defines limits to the resolving power of human eyes. The smallest state of matter is *paramaṇu* (atom) which cannot be seen. These atoms aggregate to form the world we see. The shape of atoms is spherical.[46] Two atoms combine to form a dyad. Three dyads of the same type form a triad which is large enough in size to be visible, 'as motes in a sunbeam.' Lucretius, who wrote about the connection of Democritus and India, used a similar analogy to show how invisible objects become visible.[47] The above statement of combination of atoms tells us that "atoms possess an innate propensity to aggregate. This idea is thus a forerunner of the modern concept of van der Waal forces," suggests Prof. S. K. Bose of the University of Notre Dame in Indiana, USA.[48] In his views, *Vaiśeṣika-Sūtra* "recognizes the existence of properties of the composite body that result from the manner in which the atoms are put together and organized." As an example, one can use the difference between water as liquid vs. soild. Both have the same atoms. However, the distinct properties of the bulk materials are different.

The attributes such as color, taste, smell, and touch of earth, water, fire, and air are non-eternal on account of their substrata.[49] These attributes disappear with the disappearance of their substance. Some attributes of the substrata are prone to change in its combinative (chemical action; *pākajah*) causes under the action of heat.[50] Thus, odor, flavor, and touch are attributes of atom in the *Vaiśeṣika-Sūtra* whereas Democritus denied such connections. The Greek atoms possessed only minimal properties: size, shape, and weight.[51]

The invisibility of an object, as we divide it into small parts, can simply be observed from the experience of wet clothes hung in air. The water in clothes can be seen and experienced with

[42]Marx, 1841, The Difference Between the Democritean and Epicurean Philosophy of Nature, 1841, Doctoral Thesis, Part 1, Chapter 3, online version.

[43]Sinha, 1911, p. VI; Sarsvati, 1986, p. 302; Subbarayappa, 1967. A recent paper by Karin Preisendanz assigns the second century CE to the commentary of Candrananda on *Vaiśeṣika-Sūtra*.

[44]Sarasvati, 1986, p. 295.

[45]*Vaiśeṣika-Sūtra*, 4: 1: 6; 7: 1: 9–11.

[46]*Vaiśeṣika-Sūtra*, 7: 1: 20. The statement of Kaṇāda is pretty clear. On the other hand, it was not at all possible for the ancient scholars to see an atom. How can you define a shape for something that you cannot even see? This statement is presented here to promote further scholarship on this issue.

[47]Bose, 2015. This article provides an excellent review of atomism in different cultures.

[48]Bose, 2015.

[49]*Vaiśeṣika-Sūtra*, 7: 1: 2.

[50]*Vaiśeṣika-Sūtra*, 7: 1: 6.

[51]Horne, 1960.

touch. As the sunlight and air dry the cloth, the moisture escapes from the cloth and becomes invisible. The same water drops that were visible to the human eye disappear with extended exposure to air and sunlight. Thus, smallness of an object causes this invisibility, as suggested by Kaṇāda. Continual division of matter does not lead it to disappear from existence, as it becomes invisible. It is for this reason "an absolute non-existence of all things is not possible because an atom remains in the end."[52] The *Vāyu-Purāṇa* suggested a similar ultimate division of matter, "A *paramāṇu* (atom) is very subtle. It cannot be seen by the eye. It can be imagined. What cannot be (ultimately) split in the world should be known as atom."[53]

The *Srimad-Bhāgvatam* suggested a theory of the atom that is similar to Kaṇāda's: "That which is the ultimate division of matter, that which has not gone through any change, that which is separated from others, and that which helps the perception of objects, that which remains after all is gone, - all those go under the name of *parāṃanu* (atom).[54] "Two atoms make one *aṇu* (molecule) and three *aṇu* make one *trasareṇu*. This *trasareṇu* is discovered in the line of solar light that enters into a room through a window and due to its extreme lightness such *trasareṇus* couresth the way to sky."[55]

Cyril Bailey (1871–1957), a British philologist, analyzed Kaṇāda's theory in his book, *The Greek Atomists and Epicurus*. In his view, "It is interesting to realize that at an early date Indian philosophers had arrived at an atomic explanation of the universe. The doctrines of this school were expounded in the Vaicesika Sutra [*Vaiśeṣika-Sūtras*] and interpreted by the aphorisms of Kanada [Kaṇāda]. While, like the Greek Atomists, they reached atomism through the denial of the possibility of infinite division and the assertion that indivisible particles must ultimately be reached in order to secure reality and permanence in the world, there are very considerable differences between the Indian doctrines and that of the Greeks."[56] On the relation of atoms and matter in Kaṇāda's system and its similarity to Greek atomism, Bailey continues, "Kanada [Kaṇāda] works out the idea of their combinations in a detailed system, which reminds us at once of the Pythagoreans and in some respects of modern science, holding that two atoms combined in a binary compound and three of these binaries in a triad which would be of a size to be perceptible to the senses."[57]

There are marked differences between Kaṇāda and Democritus. Atoms, as defined by Kaṇāda, are indestructible, indivisible, and have the minutest dimension. In the case of Democritus, atoms of all elements (substances) are made up of just one substance; they differ only in regards to form, dimension, position, etc. These atoms were indivisible, like Kaṇāda suggested, and had infinite shapes and, therefore, infinite types of substances.[58] The roots of Indian atomism were epistemological and based on empiricism and observation while those of Greek

[52] *Nyāya-sūtra*, 4: 2: 16.
[53] *Vāyu-Purāṇa*, 2: 39: 117.
[54] *Srimad-Bhāgvatam*, 3: 11: 1.
[55] *Srimad-Bhāgvatam*, 3: 11: 5.
[56] Bailey, 1928, p. 64.
[57] Bailey, 1928, p. 65.
[58] McDonell, 1991, p. 11–12.

atomism were ontological or *a priori*.[59] Kaṇāda gained knowledge of atoms through empiricism, intuition, logic, observation, and experimentation.

5.5 GRAVITATION AND OCEAN TIDES

The *Vaiśeṣika-Sūtra* suggested that objects fall downward when dropped as a result of gravity. "The falling of water, in absence of conjunction, is due to gravity,"[60] and "flowing results from fluidity."[61] Kaṇāda used *gurutva* word, in Sanskrit, for heaviness or gravity; *gurutvākarṣaṇ* (implying attraction of the earth) is the Hindi term for the gravitational force today.

To elaborate the gravitational properties, Kaṇāda wrote: "In absence of conjunction, falling results from gravity."[62] In the absence of any conjunction, due to gravity, only downward motion is possible: "In absence of propulsive energy generated by action, falling results from gravity."[63] Kaṇāda also explained why water should fall due to gravitational effect, even though it goes up in the form of rain-clouds during the evaporation process: "The Sun's rays cause the ascent of water, their conjunction with air."[64] Kaṇāda explains the upward motion of the objects: "Throwing upward is the joint product of gravity, volition, and conjunction."[65] An example is the upward motion of water in a tree.[66]

Al-Bīrūnī (973–1050 CE) explains the shape of the Earth and force of attraction between the Earth and human: "The difference of the times which has been remarked is one of the results of the roundity of the earth, and of its occupying the center [center] of the globe. . . the existence of men on earth is accounted for by the attraction of everything heavy toward its center [center], i.e., the middle of the world."[67] He continues, ". . . we say that the earth on all its sides is the same; all people on earth stand upright, and all heavy things fall down to earth by a law of nature, for it is the nature of the earth to attract and to keep things. . . ."[68]

Al-Bīrūnī provides an analogy of a *kadamba flower* (Fig. 4.1) to describe humans on the Earth. In this analogy, spikes are like humans. The force of gravitation allows humans to live on all sides of the Earth where nothing is up or down. He ascribed this knowledge to Varāhmihir (505–587 CE).[69] As mention in Chapter 4, Āryabhaṭa I also knew this because he used the same analogy. He said that there was no "up" or "down" side of the Earth. What this means is that, from its location in the great vastness of space, among the immense number of stars and

[59] Horne, 1960
[60] *Vaiśeṣika-Sūtra*, 5: 2: 3.
[61] *Vaiśeṣika-Sūtra*, 5: 2: 4.
[62] *Vaiśeṣika-Sūtra*, 5: 1: 7.
[63] *Vaiśeṣika-Sūtra*, 5: 1: 18.
[64] *Vaiśeṣika-Sūtra*, 5: 2: 5.
[65] *Vaiśeṣika-Sūtra*, 1: 1: 29.
[66] *Vaiśeṣika-Sūtra*, 5: 2: 7.
[67] Sachau, 1964, 1, p. 270.
[68] Sachau, 1964, vol. 1, p. 272.
[69] Sachau, 1964, vol. 1, p. 272.

other celestial bodies, it is impossible to designate an "up" or "down" side, or direction, with the Earth.

Ocean tides are noticed since antiquity. Is it a phenomenon like the overflow spill of water in cup? Do we have a similar effect in the ocean? Vālmīki, in his famous epic *Rāmāyaṇa*, connects high ocean tides to the full moon: "the roaring of the heaving ocean during the fullness of the Moon."[70] The *Viṣṇu-Purāṇa* clearly suggests that the amount of water is not increased: "In all the oceans the water remains at all times in the same quantity, and never increases or diminishes; but like the water in a pot, which, expands with heating, so the water of the ocean expand and contract with the phase of the Moon."[71]

The *Matsya-Purāṇa* also gives a similar picture, "When the Moon rises in the East, the sea begins to swell. The sea becomes less when the Moon wanes. When the sea swells, it does so with its own waters, and when it subsides, its swelling is lost in its own water . . . the store of water remains the same. The sea rises and falls, according to the phases of the Moon."[72]

This is an excellent analogy that indicates that there is an expansion in water during high tides in the ocean and the quantity of water in the ocean does not increase or decrease with a change in the phase of the Moon. Additionally, this also explains that the Hindus knew the thermal expansion of matter.[73]

Al-Bīrūnī claims that "the educated Hindus determine the daily phases of the tides by the rising and setting of the Moon, the monthly phases by the increase and waning of the Moon . . ."[74] Johannes Kepler (1571–1630 CE) is generally credited with suggesting the correlation of ocean tides with the phases of the Moon. Kepler lived about five centuries after Al-Bīrūnī and more than at least a millennium after the *Viṣṇu-Purāṇa* was written. Since the Moon and the Sun are so far away from the Earth, it was difficult for scientists of that period to comprehend the theory of ocean tides. Newton's action-at-a-distance later described its validity.

[70] Vālmīki's *Rāmāyaṇa*, 2: 6: 27.

[71] *Viṣṇu-Purāṇa*, 2: 4.

[72] *Matsya-Purāṇa*, 123: 30–34.

[73] The expansion of matter with an increase in temperature has been known to modern scientists only from the last 2–3 hundred years.

[74] Sachau, 1964, 2, p. 105.

CHAPTER 6

Chemistry

Chemistry, like medicine, evolved primarily as a science of rejuvenation among the ancient Hindus. Suśruta defined chemistry (*Rasāyana-tantra*) as the science "for the prolongation of human life, and the invigoration of memory and the vital organs of man. It deals with the recipes that enable a man to retain his manhood or youthful vigor up to a good old age, and which generally serve to make the human system immune to disease and decay."[1] Soma—the elixir of life—was produced with a knowledge of chemistry, and is mentioned in the *Ṛgveda* and *Atharvaveda*.[2] Soma is a tonic to rejuvenate a person and to slow down the aging process. The *Atharvaveda* suggests: "Invest this Soma for long life, invest him for great hearing power."[3]

Rasāyana is the rejuvenation therapy in ayurveda which was practiced to prolong life. This therapy is used to replenish the *rasa* (soup or sauce) and other *dhātus* (element) in our physical body.[4] Caraka explains the purpose of *rasāyana*: "Long life, excellent memory and intelligence, freedom from disease, a healthy glow, good complexion, a deep powerful voice, strong bodily and sensory powers, and beauty can be obtained from *rasāyana*."[5]

Caraka and Suśruta define calcination, distillation, and sublimation processes in the chemical transformation or purification of metals.[6] Caraka mentions gold, copper, lead, tin, iron, zinc, and mercury used in drugs and prescribed various ointments made from copper sulphate, iron sulphate, and sulfur, for external application in various skin diseases.[7] The oxides of copper, iron, lead, tin and zinc were also used for medicinal purposes. All these metals are native to the Indian Peninsula. Thin sheets of gold, silver, and iron were treated with salts and alkali in the preparation of drugs by Caraka.[8] The *Chāndogya-Upaniṣad* tells us of the alloys (reaction or joining) of gold with salt (borax), silver with gold, tin with silver, and copper with lead.[9]

Poisons and the drugs to counter poison were developed. The *Mahābhārata*, *Rāmāyaṇa*, and Kauṭilaya's *Arthaśāstra* mention chemical weapons (*astra*) that were used in wars. For example, Kauṭilaya provided a recipe of poison gas that was deadly: "The smoke caused by burning the powder of *śatakardama*, *uchchidiṅga* (crab), *karavira* (*nerium odorum*), *kaṭutumbi* (a kind of

[1] *Suśruta-Saṁhitā*, *Sūtrasthanam*, 1: 10.
[2] *Ṛgveda*, 10: 57: 3–4; *Ṛgveda*, 9: 62: 1; *Ṛgveda*, 9: 2: 1; *Ṛgveda*, 8: 2: 1; *Atharvaveda*, 19: 24: 3.
[3] *Atharvaveda* 29: 24: 3.
[4] *Caraka-Saṁhitā*, *Cikitsāstānam*, 1: 5
[5] *Caraka-Saṁhitā*, *Cikitsāstānam*, 1: 6–7.
[6] Biswas and Biswas, 1996 and 2001; Ray, 1948; and Bhagvat, 1933.
[7] Ray, 1956, p. 61–62.
[8] Ray, 1956, p. 62.
[9] *Chāndogya-Upaniṣad*, 4: 17: 7.

bitter gourd), and fish, together with chaff of the grains of *madana* and *kodrave* (*paspalam scrobiculatum*), or with the chaff of the seeds of *hastikarṇa* (castor oil tree) and *palāśa* (*butea frondosa*) destroys animal life as far as it is carried off by the wind."[10]

Several deadly poisons were also concocted using complicated procedures: "The smoke caused by burning the powder made of the mixture of the dung and urine of pigeons, frogs, flesh-eating animals, elephants, men, and boars, the chaff and powder of barley mixed with *kāsīsa* (green sulphate of iron), rice, the seeds of cotton, *kuṭaja* (*nerium antidysentericum*), and *kośātaki* (*luffa pentandra*), cow's urine, the root of *bhāṇdi* (*hydroeotyle asiatica*), the powder of *nimba* (*nimba meria*) *śigru* (*hyperanthera morunga*), *phaṇirjaka* (a kind of basil or *tulsī* plant), *kshibapiluka* (ripe *coreya arborea*), and *bhāṅga* (a common intoxicating drug), the skin of a snake and fish, and the powder of the nails and tusk of an elephant, all mixed with the chaff of *madana* and *kodrava* (*paspalam scrobiculatum*) or with the chaff of the seeds of *hastikaraṇa* (castor oil tree) and *palāśa* (*butea frondosa*) causes instantaneous death wherever the smoke is carried off by the wind."[11] One can notice that the concoction is fairly complicated to produce and the ingredients are not so common. The dung of pigeons, frog, and skin or a snake are not common items in most chemical preparations. However, this is how most drugs/poisons were made during the ancient and medieval periods.

The ancient Hindus used metal, minerals, gems and jewels to treat obstinate incurable diseases. Complex chemical transformations were performed before a concoction was prepared. Metals were converted into bhasms using oxidation or reduction processes. In some cases, these were transformed into biologically active nanoparticles.[12] The concept of reducing particle size for improving the effectiveness of a drug is as old as the *Caraka-Saṁhitā*. In some cases, metals are heated at a high temperature and quenched with plant extracts. In this process, flakes or metal turn into a fine nano-size powder with new chemical structure and properties. To understand if the process was achieved or not, one criterion was that the *bhasmas* should be lusterless. Al-Bīrūnī provides the accounts of alchemical practices of the Hindus.[13] "They [the Hindus] have a science similar to alchemy which is quite peculiar to them. They call it *Rasāyana*, a word composed with *rasa*, i.e., gold.[14] It means an art which is restricted to certain operations, drugs, and compound medicines, most of which are taken from plants. Its principles restore the health of those who are ill beyond hope, and give back youth to fading old age, so that people become again what they were in the age near puberty; white hair becomes black again, the keenness of the senses is restored as well as the capacity for juvenile agility, and even for cohabitation, and the life of people in this world is even extended to a long period."[15]

[10] *Arthaśāstra*, 411; Book 14, Chapter 1; Shamasastri, 1960. p. 442.

[11] *Arthaśāstra*, 411; Book 14, Chapter 1; Shamasastri, 1960. p. 442–443.

[12] Chaudhary and Singh, 2010; Sarkar and Chaudhary, 2010

[13] Sachau, vol. 1, p. 187–193.

[14] Al-Bīrūnī is mistaken in translating *rasa* as gold. Since gold is used in some recipes in *Rasāyana*, it may have caused him to make this mistake.

[15] Sachau, vol. 1, p. 188–189.

"In India the earliest allusions to alchemical ideas appear in the Atharva Veda (*Atharvaveda*), where mention is made of the gold which is born from fire," writes C. J. Thompson, in his book *The Lure and Romance of Alchemy*.[16] The *Ṛgveda* mentions *surā* as an intoxicating drink that was used along with soma. Today, *surā* is a term used in India for alcoholic drinks. The *Arthaśāstra* of Kauṭilaya gives recipes for fermented alcoholic drinks made from rice (similar to Sake, a popular Japanese drink), sugarcane (similar to rum), grapes (similar to wine), and various spices. Suśruta wrote a complete chapter on the elixirs for rejuvenation of humans and suggested recipes for people to become immune to disease and decay.[17]

Cyavanaprāśa is a very popular tonic for rejuvenation in India. The name has stemmed from the name of Cayavana *ṛṣi* (seers) who rejuvenated his body with the help of Aśvin Kumāras, two Vedic doctors. This is mentioned at several places in the *Ṛgveda*.[18] People in northern India take *Cyavanaprāśa* as a precautionary measure particularly in the winter season against the common cold and cold-related symptoms. The use of *Cyavanaprāśa* is fairly ancient as Caraka has referred to it for rejuvenation.[19]

Caraka provided several recipes for making iron tonics in his *Caraka-Saṁhitā*. The process of making these elixirs is known as the *killing of metal* in Sanskrit, and the final product is called *bhasma*. To accomplish the killing of iron, Caraka recommended the use of fine thin plates of iron with *āmlā* (a fruit) extract and honey for one year in an underground pot; it reduces iron to a ferrous compound. Also, Caraka mentioned the conversion of yellow gold into a red colloidal form using plant extract, before being used as a tonic. Many of these reactions were done to make the metals nontoxic. Suśruta devoted one whole chapter on the use of alkalis (*kṣāra*) in various diseases and noticed its corroding, digestive, mucous destroying, and virile potency destroying properties.[20]

6.1 MINING AND METALLURGY

The ancient Hindus excelled in mining and metallurgical processes. In Persia, due to the high quality of the so-called Damascus steel that was originally produced in Hind, steel was called "*foulade Hind*," indicating the steel of India. Similarly, after the conquest of Poros, Alexander the Great received steel as the precious gift that the Greeks did not have. (Read Section 6.2.) Aristotle, in his book, *On Marvelous Things Heard*, writes that Indian-copper is known to be good and is indistinguishable from gold.[21] This mention was made perhaps to acknowledge the high quality of the Indian bronze made from copper that was "indistinguishable from gold" due to the mixing of zinc and other metals or minerals.

[16] Thompson, 1932, p. 54.
[17] *Suśruta-Saṁhitā*, *Cikitsāstānam*, 27: 1.
[18] *Ṛgveda*, 5: 74: 5 and 6; *Ṛgveda*, 7: 71: 5.
[19] *Caraka-Saṁhitā*, *Cikitsāstānam*, 1: 74.
[20] *Suśruta-Saṁhitā*, *Sūtrasthānam*, Chapter 11.
[21] Aristotle, *On Marvelous Things Heard*, 49, part of Aristotle's *Minor Work*.

The smelting operation is defined in *Ṛgveda*[22] where ore was dumped into fire and bellows were used to fan the fire. A good example of its existence is the theory of creation, as shared in Chapter 4, provided in *Ṛgveda*, where the analogy of a smelting device was used to describe the creation of the universe. The *Atharvaveda* tells us that the chest of the earth contains gold, indicating gold mines and mining operation.[23] Most ancient metals such as iron, copper, gold, silver and lead were naturally available in various parts of India.[24] The remains of Mohenjo-daro and Harappa contain metallic sheets and pottery that are inscribed. Bronze, copper, iron, lead, gold, and silver were used to make axes, daggers, knives, spears, arrow heads, swords, drills, metal mirrors, eating and cooking utensils, and storage utensils. The 10 centimeter high Mohenjo-daro's dancing girl is made of copper with armful of bangles (Figure 6.1). Out of the 324 objects from these archaeological sites that have been analyzed, about 184 objects are of pure copper[25] and four archaeological copper processing kilns are discovered at Harappa, Lothal, and Mohenjo-daro's sites.[26]

The timber from an old mine in Hutti, Karnataka was collected at a depth of about 200 meters and was carbon dated. It was concluded that the mining process was carried out in this mine around 4th century BCE.[27] Ktesias, a Greek traveler who lived in Persia during the 5th century BCE, mentions of the vast amount of gold mined from "high-towering mountains."[28] He also mentions a congealing processing to get quality gold.[29] Several ancient Greek historians, such as Herodotus, Ktesias, Arrian, and Megasthenes, mentioned the use of metals in India.[30]

The chemical analysis of the brass artifacts from Taxila (third-fourth century BCE) reveal that 35–40 percent zinc content in these artifacts, giving a golden appearance to them. This was achieved by mixing zinc with copper. It is quite natural to assume that a metallurgical process was practiced to get pure zinc. Zinc is found as sphalerite, a sulphide of zinc. This ore is first converted into an oxide in a combustion process. This is further converted into zinc in a reduction process which is quite cumbersome. However, the smelting technicians found an innovative procedure to get pure zinc. In contrast, such extraction methods were practiced in Europe only around the sixteenth century.[31]

Zinc was smelted in a downward distillation process where the zinc vapors were swiftly cooled down to avoid reduction process that happens at a slightly higher temperature. The main concern is that the zinc oxide needs a minimum temperature of 1150 degree Celsius for the

[22] *Ṛgveda*, 9: 112.
[23] *Atharvaveda*, 12:1:6.
[24] Agarwal, 2000; Biswas and Biswas, 1996.
[25] Lahiri, 1995.
[26] Agrawal, 2000, p. 40.
[27] Radhakrishna and Curtis, 1991, p. 23–24.
[28] McCrindle, 1973, p. 16, 17.
[29] McCrindle, 1973, p. 68.
[30] Bigwood, 1995.
[31] Subbarayappa, 2013 p. 299.

Figure 6.1: The dancing girl of Mohenjo-daro. (Taken from Wikimedia).

reduction of oxide while the boiling point of zinc is 900 degree Celsius. Thus, it vaporizes and escapes. The excavated furnaces at Zawar, in Udaipur district, Rajasthan, have two chambers, top and bottom, separated by a thick perforated brick plate. The top chamber is sealed from the top, forcing vapor to go to the lower chamber which was filled with water. Water reduced the temperature and solidified zinc. Even today, this area is known for mining operations and Hindustan Zinc Ltd. is still in operation there.[32]

The copper statue of Buddha in Sultanganj (Figure 6.2), Bihar, is over 2.3 m high, 1 m wide, and weighs over 500 kg. It was found in 1864 in the North Indian town of Sultanganj, Bhagalpur district, Bihar, by an East India Company engineer. Today, the statue is housed in Birmingham Museum and Art Gallery, Birmingham, England. The statue was made using a technique which is popularly known as 'lost wax technique'. It is a method of metal casting in which a molten metal is poured into a mold, mostly of clay, that is created by means of a wax model first. Once the soft clay solidifies, the whole assembly is baked, melting the wax and extracted. This clay mold is then used for metal casting.

The Painted Grey Ware of the Gangetic Valley were manufactured around 600 BCE.[33] Two glass bangles found at Hastinapur, near New Delhi, are from the 1100–800 BCE period. The bangles are made from soda-lime-silicate glass with ornamental coloring from iron. Similar bangles of green color were also found elsewhere in northern India.[34] Glass bangles are the ornamental jewelry that are worn by Hindu ladies even today. Beads of green glass with cuprous oxide coloring were found in Taxila belonging to the 700–600 BCE period.[35]

Kauṭilya realized the importance of mining to the state economy. He suggested that "mines are the source of treasury; from treasury comes the power of government."[36] He described various metallic ores and mining operations along with the processes of distillation, refinement, and quality control. Kauṭilya even defined the duties of the Director of mining: "must possess the knowledge of the science dealing with copper and other minerals (*sulbadhātu-śāstra*), experienced in the art of distillation and condensation of mercury (*rasapāka*) and of testing gems, aided by experts in mineralogy and equipped with mining laborers and necessary instruments, the Director of mines shall examine mines which, on account of their containing mineral excrement, crucibles, charcoal, and ashes, may appear to have been once exploited or which may be newly discovered on plains or mountain slopes possessing mineral ores, the richness of which can be ascertained by weight, depth of color, piercing smell, and taste."[37]

Kauṭilaya provided clues for finding good locations for mining: observe soil, stones, or water that appears to be mixed with metal and the color is bright or if the object is heavy or with strong smell, this indicates the possibility of a mine nearby.[38] He provides techniques to locate glass, gold, silver, iron, and copper mines.[39] Mining operations were leased to private investors at a fixed rate.[40] According to him, the head of mining operations must know metal-

[32]Subbarayappa, 2013, p. 300; Srinivasan, 2016.
[33]Dikshit, 1969, p. 3.
[34]Dikshit, 1969, p. 3.
[35]Dikshit, 1969, p. 4.
[36]*Arthaśāstra*, 85; Book 2, Chapter 12; Shamasastri, 1960. p. 89.
[37]*Arthaśāstra*, 82; Book 2, Chapter 12; Shamasastri, 1960. p. 83.
[38]*Arthaśāstra*, 82; Book 2, Chapter 12; Shamasastri, 1960. p. 84.
[39]*Arthaśāstra*, 82; Book 2, Chapter 12; Shamasastri, 1960. p. 84.
[40]*Arthaśāstra*, 83; Book 2, Chapter 12; Shamasastri, 1960. p. 86.

Figure 6.2: Statue of Buddha from Sultanganj. (Taken from Wikimedia.)

lurgy, chemistry, and the refinery processes.[41] This tells us that metallurgy was an established discipline during the third century BCE in India. Kauṭilaya defined the ethical rules of conduct for goldsmiths. Severe penalties were prescribed for goldsmiths who fraudulently adulterated gold. Any mixing of inexpensive impurities of tin, copper, and brass during the melting process of gold and silver was considered a crime and the criminals were prosecuted. Thieves of mineral products were punished with a financial penalty that was eight times the value of the stolen products.[42]

[41] *Arthaśāstra*, 81; Book 2, Chapter 12; Shamasastri, 1960. p. 83.
[42] *Arthaśāstra*, 83; Book 2, Chapter 12; Shamasastri, 1960. p. 86.

Kauṭilaya described gold, silver, arsenic, copper, lead, tin, and iron ores and their purification processes.[43] Coins, as currency, were used in business transactions during his period. Compositions of different coins are defined in Kauṭilaya's book,[44] and rules of punishment are defined for people making counterfeit coins.[45] Kauṭilaya also explained the methods to catch manufacturers of counterfeit coins[46] and defined a penalty for the coin-examiner to accept a counterfeit coin into the treasury.[47]

Sixteen gold standards based on the percentage of gold content in a specimen, that is similar to the current "carat" system, were defined by Kauṭilaya. Copper was mixed with gold to form an alloy of different carat standards.[48] Several gold alloys were defined by Kauṭilaya that were blue, red, white, yellow, and parrot (green) in colors. This was achieved by the processing of gold with rock salt, lead, copper, silver, mercury, etc. The processes of chemical reactions were well-defined with exact proportions of each element or compound.[49] Gold-plating and other metal-plating procedures are defined using amalgams, heating processes, rock salt, mica, wax, etc.[50] Kauṭilaya also defined color, weight, characteristics, hammering, cutting, scratching, and rubbing as tools to test precious stones.[51]

6.1.1 THE IRON PILLAR OF NEW DELHI

The Iron Pillar near Qutab-Minar in New Delhi is a testimonial to the metal forging skills of the ancient Hindus (Figure 6.3). The pillar marks the renunciation of kingly duties and the beginning of aesthetic life for King Candra. We know of this from the inscription on the pillar. The inscription is dated between 400–450 CE and the inscribed letters have minimal corrosion despite 1600 years of weathering in the open air.[52] Air, heat, and heavy rains of Northern India have not caused significant rusting of the pillar, even with high heat and humid weather from July to September. The pillar is indisputably a long standing permanent record of the excellent metallurgical skills and the engineering skills of the ancient Hindus.

King Candra is most likely the famous King Candragupta Vikramāditya II (375–413 CE) and the current site of the pillar was chosen by Tomar King Anaṅgpāla who erected it on the site of a temple. In 1739 CE, Nadir Shah occupied the city and decided to break the pillar using cannons and failed. Several spots where the cannon balls hit the pillar are still marked.[53]

[43] *Arthaśāstra*, 93; Book 2, Chapter 14; Shamasastri, 1960. p. 98.
[44] *Arthaśāstra*, 84; Book 2, Chapter 2; Shamasastri, 1960. p. 86–87.
[45] *Arthaśāstra*, 59; Book 2, Chapter 5; Shamasastri, 1960. p. 57, 58.
[46] *Arthaśāstra*, 212; Boook 4, Chapter 4; Shamasastri, 1960. p. 239.
[47] *Arthaśāstra*, 203; Book 4, Chapter 1; Shamasastri, 1960. p. 230.
[48] *Arthaśāstra*, 86; Book 2, Chapter 13; Shamasastri, 1960. p. 90.
[49] *Arthaśāstra*, 88; Book 2, Chapter 13; Shamasastri, 1960. p. 92.
[50] *Arthaśāstra*, 92; Book 2, Chapter 14; Shamasastri, 1960. p. 97.
[51] *Arthaśāstra*, 92; Book 2, Chapter 14; Shamasastri, 1960. p. 97.
[52] Balasubramaniam, 2001.
[53] Balasubramaniam, 2001; Balasubramaniam, Prabhakar and Shanker, 2009.

Figure 6.3: The Iron Pillar of New Delhi. (Taken from Wikimedia.)

The pillar is about 7.16 m long (23 feet, 6 inches) with a 42.4 cm (16.4 inches) diameter near the bottom and about 30.1 cm (11.8 inches) at the top. It weighs over six tons. The pillar is a solid body with a mechanical yield strength of 23.5 tons per square inch and ultimate tensile strength of 23.9 tons per square inch. It is made of wrought iron. There is no other pillar from the early medieval period of that size anywhere else in the world.[54]

The composition of the wrought iron is as follows:[55] carbon 0.15%, silicon 0.05%, sulfur 0.005%, manganese 0.05%, copper 0.03%, nickel 0.05%, nitrogen 0.02%, phosphorous 0.25%, and 99.4% pure iron.

Iron of such purity is not naturally available in Indian mines. This composition is a strong indication of an iron refining process in the making of this pillar. The pillar has a coating of a thin protective layer of Fe_3O_4 using salts and quenching. The excavation of the buried portion revealed that the base of the pillar was covered by a sheet of lead, about 3 mm in thickness. This absence of rusting in not unique to the Iron Pillar. The iron beams of the Sun Temple in coastal

[54]Lal, B. B., On the Metallurgy of the Meharauli Iron Pillar, in the book by Joshi and Gupta, 1989, p. 25 and 26; Balasubramaniam, 2001.

[55]Biswas and Biswas, 1996, vol. 1, p. 394.

Orissa and the Iron Pillar of Dhar (Madya Pradesh), both in relative higher humidity regions, do not show much rusting either. This is perhaps due to a higher phosphorus content of the iron.

The Sun Temple of Koṇārka, a ninth century construction, in Orissa has 29 iron beams of various dimensions used in its construction. The largest beam is 10.5 meter (35 feet) in length and about 17 cm (about 7 inches) in width with a square cross-section and is about 2669 kg (6000 lbs.) in weight.[56] Iron nails are used to connect stone pieces. The composition of the iron beams and nails is similar to those of the famed Iron Pillar of Delhi. Similarly, the iron pillar of Dhar is about thousand years old and has little or no rusting.[57] The pillar is 7.3 tons in mass and 42 feet long. Dhar is an old capital of Malwa which is about 60 km from the well known city of Indore. The pillar was possibly constructed by King Bhoja (1010–1053 CE) and currently lies in three parts in front of Lāṭ mosque in Dhar. King Bhoja was well versed in metallurgy and wrote a book on metallurgical processes and metal weapons, called *Yuktikalpataru*. Bahadur Shah, a Muslim king, captured the region and decided to move the erected pillar to Gujrat. In the process of digging ground to take out the pillar, it fell down and was broken into pieces, as we know from the memoirs of Emperor Jahāngir. The pillar has been weathering the monsoons of India for over a thousand years and has little rusting. The surface of the pillar is coated with a thin optically dull layer and on top of it is another thick layer of optically bright material. Professor R. Balasubramaniam of the Indian Institute of Technology, Kanpur, India has compiled a nice summary of the history of the pillar and its chemical constitution.[58]

Recent researches have figured out the process on why the pillars do not rust. The main reason is the mixing of phosphorus with iron. Once the surface of iron is rusted, it becomes porous and allows phosphorus to react with other chemical compounds which reduces to phosphoric acid. This acid interacts with iron to form dihydrogen phosphate. The chemical reactions are as follows:[59]

$$2H_3PO_4 + Fe \rightarrow Fe(H_2PO_4)_2 + H_2$$

$$2H_3PO_4 + FeO \rightarrow Fe(H_2PO_4)_2 + H_2O$$

This further dissociates into two forms:

$$3Fe(H_2PO_4)_2 \rightarrow Fe_3(PO_4)_2 + 4H_3PO_4$$

$$Fe(H_2PO_4)_2 \rightarrow FeHPO_4 + H_3PO_4$$

Both of these phosphates are amorphous and insoluble in water. These amorphous phosphates reorganize themselves into crystalline ferric phosphate, and drastically reduces the poros-

[56] Ray, 1956, p. 212.
[57] Balasubramaniam and Kumar, 2003.
[58] Balasubramaniam, 2002.
[59] Balasubramaniam, 2001.

ity of the surface. This large reduction in porosity reduces any further rusting of the iron.[60] Thus, it is the phosphorus content of the pillar that does the trick, making the pillar to be rust free. Can we use the same technology in making car bodies which are prone to rusting in cold regions where salt is used on roads?

6.2 *WOOTZ* OR DAMASCUS STEEL

The word steel comes from the Old High German (German language, around 11th century CE) word *stahal* which is related to the Sanskrit word *stakati*, meaning "it resists"[61] or "strike against." Sword-making was a popular use of this steel due to its hardness. Steel is an alloy of iron containing 0.10 to 1.5% carbon in the form of cementite (Fe_3C).[62] The properties of steel vary greatly with a minor change in carbon content, along with other elements. Metals such as manganese, silicon, chromium, molybdenum, vanadium, or nickel are purposely mixed in the process depending on the desired outcome. For example, stainless steel has approximately 12% or more chromium content.

Steel was prepared and used in India for various purposes from the ancient period.[63] Ktesias, who was at the court of Persia during the 5th century BCE, mentioned the two high quality Indian steel swords that were presented to him. One sword was presented by the King of Persia and the other by King's mother, Parysatis.[64] Nearchus (fl. 360–300 BCE), an officer in the army of Alexander the Great, mentions that the Indians generally carried a broad three cubit long sword with them.[65]

King Poros, a king from Punjab, lost a major battle with Alexander the Great and was imprisoned. He was brought to the court and Alexander asked him how he should be treated now. Poros replied: like a king. This surprised Alexander and the ensuing conversation between them made him realize the futility of wars. Alexander decided to release Poros and gave his kingdom back to him. This was a highly unusual move from Alexander. Usually defeated kings wer slaughtered or imprisoned. Poros was grateful to receive a gift of life and wanted to show his gratitude to Alexander the Great. Porus wanted to gift something that was precious and Alexander did not have—more precious than gold, gems, or spices. He opted to give 100 *talents* (6000 pounds)[66] of steel as a precious gift to Alexander, as we know from the accounts of Quintus Curtius (9: 8: 1), a Roman historian during the first century CE who wrote a biography of Alexander the Great. Steel was called *ferrum candidum*, meaning white iron, by Curtius.[67]

[60] Balasubramaniam, 2001.
[61] Le Coze, 2003.
[62] Bhardwaj, 1979, p. 159.
[63] Prakash and Igaki, 1984.
[64] *Fragments*, 1: 4; McCrindle, 1973, p. 9.
[65] Bigwood, 1995.
[66] Casson, 1989, p. 114.
[67] Casson, 1989, p. 114.

The Periplus Maris Erythraei text, written around 40–70 CE,[68] provides information on the importation of steel to the Roman empire. It was subjected to a custom duty under Marcus Aurelius (121–180 CE) and his son Commodus (161–192 CE).[69] During the period of Emperior Justinian (482–565 CE) of Rome, the *Digest of the Roman Law*, (39, 15, 5–7) was compiled to run the state. Indian iron (*Ferrum Indicum*) is in the list of objects that were subject to import duty.[70] Similarly, iron was taxed and traded in Alexandria, Egypt in the early first century CE and was called *koos*. Steel is called *wuz* in the Gujrati language, and *Wooku* (or *ukku*) in the Kannada language, a prominent language in South India.[71] Perhaps this explains the origin of *wootz*, the Syrian word for steel. This word was new to the Syriac literature and appeared late in chronology.

The Arabs called steel *Hundwáníy*, meaning Indian.[72] This word perhaps evolved into *andanic* or *ondanique* for swords and mirrors, used by the medieval writers. This also led to the words *alhinde* [or al-Hind, meaning India] and *alinde* [for steel mirror] in Spain.[73] The best steel in Persia was called *foulade Hind*, meaning steel of India. Another kind of steel, *jawābae Hind*, meaning a Hindu answer, was also popular because it could cut a steel sword.[74]

Steel was called *wootz* in India and was traded in the form of castings (cakes) of the size of ice-hockey pucks.[75] Persians made swords from *wootz* and these swords were later erroneously known as Damascus swords.[76] As is the case with the Arabic numerals, the Europeans learned about steel-making in the Middle East during the War of the Crusades in 1192 CE.[77] By noticing the remarkable hardness and strength, they became interested in knowing the secrets of making ultra-high carbon steel. The European did not know at that time that the manufacturing process had originated in India.

Archaeological sites in the Periyar district of Tamil Nadu, which date back to about 250 BCE, provide indications of crucibles used to mix iron and carbon for the steel-making process.[78] Varāhmihir, during the sixth century CE, wrote a chapter on swords (*khaḍgalakṣanam*) in his *Bṛhat-Saṁhitā* (50: 23–26), and provided a recipe for the hardening of steel. His processes used chemical techniques as well as heating and quenching. Swords treated with these processes did not "break on stones" or become "blunt on other instruments," as reported by Varāhmihir.[79]

Steel production was very much limited due to the high consumption of fuel and the required high temperature for melting. The situation prevailed until 1850 CE, when high tem-

[68]Casson, 1989, p. 7.
[69]Casson, 1989, p. 114.
[70]Schoff, 1915.
[71]Agarwal, 2000, p. 198; Biswas and Biswas, 1996, vol.1, p. 121, Le Coze, 2003.
[72]Schoff, 1915.
[73]Schoff, 1915.
[74]Royle, 1837, p. 47.
[75]Sherby and Wadsworth, 2001
[76]Sherby and Wadsworth, 2001
[77]Joshi, Narayan, R., Tough Steel of Ancient India, p. 293, in the book by Sharma and Ghose, 1998.
[78]Agarwal, 2000, 197; Prakash and Igaki, 1984.
[79]Biswas and Biswas, 1996, vol. 1, p. 276.

perature furnace technology improved. Thus, the use of steel was largely for making blades for knives, daggers, and swords. With proper processing, these steels can be made to a strength that is about five times greater than that of the strongest wrought iron.[80] Damascus steel has an attractive swirling surface pattern that is an outcome of the cooling process. The patterns result from the alignment of the Fe_3C particles that form on the surface during the cooling process.[81] Thus, Damascus swords became famous for their hardness and could absorb blows in combat without breaking. It did not fail a Middle Age warrior during combat.

Making *wootz* steel is complex process as even minute impurities change the outcome. Temperature and the cooling period also affect the quality of steel. Giambatlista della Porta (1535–1615), an Italian scholar from Naple, in 1589 CE wrote on the importance of temperature in treating *wootz*, and suggested to avoid "too much heat."[82] Oleg D. Sherby and Jeffrey Wadsworth, two researchers from Stanford University, figured out that a slow equilibrium cooling is the best way to produce quality steel. When iron and carbon (1.3–1.9%) are heated to 1200°C, they reach a molten state and the slow cooling allows the carbon to diffuse through the iron to form white cementite patterns that result from alignment of the Fe_3C (cementite) particles. With polishing, the Fe_3C particles appear white in the near black steel matrix. The carbide particles serve the role of strengthening without making the metal brittle.[83] The blade was hardened by heating it to 727°C which allows a change in the crystal structure. Iron molecules that were distributed as body-centered ferrite begin to form a face-centered lattice. The blade was then quenched in water.[84] If the heating was done above 800°C before quenching, it made the metal brittle.

Michael Faraday (1791–1867), a fellow of the Royal Society and discoverer of many electromagnetic discoveries, tried to duplicate Damascus steel and incorrectly concluded that aluminum oxide and silica additions contributed to the properties of the steel. Faraday and Stodart also attempted to make steel by alloying nickel and noble metals like platinum and silver. All their efforts were of no avail. Thus, the steel-making technology of ancient India alluded even great European scientists till the nineteenth-century.[85].

6.3 FERMENTATION

The *Ṛgveda* mentions fermentation of barley, a key ingredient in beer: "A mixture of a thick juice of soma with barley powder."[86] In another place, there is a mention of fermented alcoholic drink which took about 15 days of processing. "Fifteenth day old highly intoxicating soma,"[87] which

[80] Sherby and Woodworth, 2001
[81] For more information, please read, Sache, 1994; Figiel, 1991.
[82] Smith, 1982.
[83] Sherby and Wadsworth, 1985.
[84] Sherby and Wadsworth, 1985.
[85] Faraday, 1819; Stodart and Faraday, 1822; Day, 1995
[86] *Ṛgveda* IX: 68: 4.
[87] *Ṛgveda* X: 27: 2.

probably refers to the broth fermented in the vat for 15 days. The *Yajurveda-Śukla* tells us about various kinds of alcoholic drinks:[88] "Âtithya's sign is Mâsara, the Gharm's symbol *Nagnahu*. Three nights with *Surâ* poured, this is the symbol of the *Upasads*. Emblem of purchased Soma is Parisrut, foaming drink effused." Elsewhere in the same book, the fermentation process is mentioned. "Like shuttle through the loom the steady ferment mixes the red juice with the foaming spirit."[89] Here *Nagnahu* is root of a plant that is used as yeast, Parisrut is a kind of beer, and *Surâ* is another word for liquor.

The *Chāndogya-Upaniṣad* tells us that drinking liquor on a regular basis is as bad as stealing gold, killing a Brahmin, or having an affair with your teacher's wife.[90] Elsewhere, five extremely wealthy and immensely learned householders decided to examine the following two questions: What is our self (*ātman*)? What is *brahman*? They decided to visit Aśvapati Kaikeya, a highly learned scholar, to teach them about these two questions. Aśvapati Kaikeya tells his visitors that "no one drinks" in his kingdom, indicating the presence of alcoholic beverages.[91]

Caraka mentions some 84 different kinds of alcoholic liquors.[92] For the fermentation, Caraka mentions the following sources of sugar: sugarcane juice, guḍa (jaggery), molasses, honey, coconut water, sweet palmyra sap and *mahua* flowers. Some sweet fruits such as grape, date, mango, banana, apricot, jackfruit, rose-apple, *jāmun*, pomegranate, *kādamba*, *bilva*, etc. are also used in the fermentation. Similarly, rice and barley were used from the grain category for fermentation. By the time of Kauṭilaya, the superintendent of liquor was designated to supervise this industry. The manufacturing of liquor and controlled liquor traffic in and out of village boundaries were monitored.[93] Liquor shops were required to have beds and seats and the rooms were filled with the scents of flowers. Some recipes for liquor making are also provided: "Medaka is manufactured with one droṇa (measure of capacity, volume) of water, half an āḍhaka (unit of mass) of rice, and three prasthas (mass) of *kiṇva* (ferment)."[94]

Some ferments were used for medicinal purposes. Kauṭilaya suggested that patients should learn the preparation of these *ariṣṭa* (fermented and distilled liquor and medicine) from physicians. He has even provided several recipes: "one hundred pala of *kapittha* (*Feronia Elephantum*), 500 pala (mass) of sugar, and one prastha (mass) of honey form *āsava*."[95] Some hard liquors with rice and lentils were also suggested: "one droṇa of either boiled or unboiled paste of *māsa* (*Phraseolus Radiatus*), three parts more of rice, and one karṣa of *morata* (*Alangium salviifolium*) and the like form *kiṇva* (ferment)."[96] Additives were mixed in order to improve the taste or appearance

[88] *Yajurveda-Śukla*, 19: 13–15.

[89] *Yajurveda-Śukla*, 19: 83.

[90] "A man who steals gold, drinks liquor, and kill a Brahmin; A man who fornicates with his teacher's wife—these four will fall." *Chāndogya-Upaniṣad*, 5: 10: 9.

[91] *Chāndogya-Upaniṣad*, 5: 11: 5.

[92] Achaya, 1991.

[93] *Arthaśāstra*, 119; Shamasastri, 1960. p. 131.

[94] *Arthaśāstra*, 120; Book 3, Chapter 25; Shamasastri, 1960. p. 133.

[95] *Arthaśāstra*, 120; Book 2, Chapter 25; Shamasastri, 1960. p. 132.

[96] *Arthaśāstra*, 120; Shamasastri, 1960. p. 133.

of liquors. These additives were used sweeteners, spices, and astringents.[97] The variety listed in *Caraka-Saṁhitā* and *Arthaśāstra* is comparable or even better than the variety that is present in the market today.

[97]For more information, read Achaya, 1991; Prakash, 1961; and Singh *et al.*, 2010.

CHAPTER 7

Biology

In the *Caraka-Saṁhitā*, an early Hindu treatise on medicine, biology was considered the most important of all sciences. "The science relating to life is regarded by the philosophers as the most meritorious of all the sciences because it teaches mankind what constitutes their good in both the worlds [spiritual and physical]."[1] Good health helps people to fulfill their four purposes in life: *dharma* (duty), *artha* (prosperity), *kāma* (sensuality), and *mokṣa* (liberation).[2]

The ancient Hindus systematically studied various life forms and noticed inter-dependencies and commonness in them. It was due to these commonnesses between plants and animals that led Suśruta to suggest that new medical practitioners should dissect plants first before they performed any dissection on animals or humans. About 739 plants and over 250 animals are mentioned in the ancient literature of the Hindus.[3] The 24th chapter of *Yajurveda* has a large variety of birds, animals, and snakes mentioned. Plants and animals are characterized with respect to their utility, habitat, or some other special features: Four-footed, reptiles, claws, born from an embryonic sac, born from an egg, born from sprouts (plants), or domesticated.[4] The botanical world was divided into grasses, creepers, shrubs, herbs, and trees.[5] The *Ṛgveda* mentions the heart, lung, stomach, and kidneys. The *Atharvaveda*[6] refers to the heart as a "lotus with nine gates," a correct description of heart when held with its apex upwards. From this top view, one can observe nine openings in the heart: three in right atrium, four in the left atrium, and one in each of the right and left ventricles.[7] Similarly, perhaps there is a mention of blood circulation in the Atharvaveda:[8] "Who stored in him floods moving in all diverse directions and formed to flow in rivers pink, rosy red, and coppery dark running in all ways in a man, upward and downward."

[1] *Caraka-Saṁhitā, Sūtrasthānam*, 1: 43.
[2] *Caraka-Saṁhitā, Sūtrasthānam*, 1: 15.
[3] Kapil, 1970.
[4] Smith, 1991.
[5] Kapil, 1970; Smith, 1991.
[6] *Atharvaveda*, 10: 8: 43.
[7] Narayana, 1995; Rajgopal et al., 2002.
[8] *Atharvaveda*, 10: 2: 11.

7.1 SACRED RIVERS AND MOUNTAINS: ECOLOGICAL PERSPECTIVES

Ecology (*paryāvaraṇa*) primarily deals with the relationships of organisms with their environment. The Hindus strongly believe in the mighty Earth as a living system that has several billion years of experience in developing processes that are sustainable and cyclic. It is the respect for the ecosystem that led the ancient Hindus to promote vegetarianism. Trees, mountains, rivers, and animals were worshiped by the ancient Hindus. Cutting trees, dumping waste products in a river, and killing animals are considered as sins. It is a common phrase in India that the God lives in every particle of the universe. (*kaṇa kaṇa mei Bhagwān hai*). The *Bhagavad-Gītā*[9] suggests the omnipresence of God. Therefore, in the worldview of the Hindus, the ecology should be preserved and respected. The *Manu-Smṛti* tells us that that all plants and animals have specific functions and we must protect them: "Brahma, the God of Creation, has created all the plants and animals with specific characteristics and functions, and none should disturb these creatures."[10]

The ancient Hindus considered nature (*prakṛti*) as a living sacred entity. Nature is not a property that is owned by the humans for their use; on the contrary, humans are a part of nature. Nature is represented by its five forces: *ākāśa* (space or sky), *vāyu* (air), *teja* (fire), *ap* (water) and *pṛthivī* (earth)—popularly known as *pañcā-bhūta* (five elements). All animate and inanimate objects in nature are made up of these five elements, including humans and animals. The ancient Hindus realized that they could neither control nature nor overpower its order; they just tried to live with it in a manner that was based on respect and appreciation for the natural forces and order. Rivers, mountains, and oceans were treated with reverence. It was not ethical to dump waste into a river or deplete a mountain of its trees. The ecosystem connects living beings with inanimate objects. For example, a little misuse of water, an inanimate object, has dire consequences to life forms that drink it.

The geographical region of the Indus Valley and the surrounding area was considered sacred by the ancient Hindus. It was only natural for them to preserve the ecology of the region. The Earth is worshiped as mother; it is considered to be a *devī* (Goddess) and has numerous names: *Bhūmi*, *Pṛthivī*, *Vasudha*, and *Vasundharā*. Even today, India is personified as *Bhārat-mātā* (Mother India). The Earth is considered as mother and the inhabitants as her children, as suggested in the *Atharvaveda*,[11], and supports people of all races and remains fertile, arable and nourisher of all.[12] This chapter of the *Atharvaveda* does not discriminate between "us" and "them;" it includes everyone. Another verse in the same book suggests that the Earth caters to

[9] *Bhagavad-Gītā*, 7: 19; 13:13.
[10] *Manu-Smṛti*, 1: 28: 39–35.
[11] *Atharvaveda*, 12: 1: 12.
[12] *Atharvaveda*, 12: 1: 11

people of different religions and languages.[13]. A request is made for Mother Earth to provide medicines, medicinal plants and prosperity,[14] and prestige.[15]

Every day, hundreds of thousands of people visit Ganges [Gaṅgā] or Yamunā rivers and worship them. These rivers are considered to be purifiers of sins, and bathing in the sacred waters is a popular activity which is also associated with a prayer. In the evenings, group *ārati* (prayer) is performed where people worship God by worshiping the river. As David Gosling, a nuclear physicist who has studied South Asian ecology, suggests that the Hindu traditions serve as an important model in "raising social and environmental awareness, underscoring the continuities between past and present and their possible transformations within an environmental paradigm."[16]

7.2 SACRED *TULSĪ* AND SACRED COW

Trees and plants are highly efficient chemical and pharmaceutical factories. Human survival is dependent on the quality of the foods provided by trees and plants. It is only natural to respect plant life and worship them. The respect that the Hindus bestow on plants can be observed by noticing stone deities placed under many trees with flags and steamers adorning the branches. The mythological *kalpavṛkṣa*, a tree that fulfills all human desires, is well known in Hindu tradition. In their popular belief, each tree has a *vṛkṣa-devatā*, or tree-deity, who is worshiped by the Hindus with prayers and offerings of water, flower, sweets, and encircled by sacred thread. The sacred thread symbolizes the wishes of the praying person. For example, in the *Mahābhārata*, Sukra says: "Rubbed with the astringent powder of the hanging roots of the Banyan tree (*Ficus bengalensis*) and anointed with the oil of *priyango* (*panicum italicum*), one should eat the *shashlika* paddy mixed with milk, By so doing one gets cleansed of all sins."[17] The *Padma-Purāṇa* advocated that planting trees was a simple way to reach *nirvāṇa* (liberation), to escape the cycle of birth and death.[18]

Since the Hindus realized the common connection between plants, animals and humans, and the importance of trees for human life, it became a religious code to preserve plant life. Aśoka (reigned 272–232 BCE) issued a decree against the burning of forests.[19] It was a hunting practice during the period to set fire to grass and trees. As the animals got scared by the fire and ran away to protect themselves, they were trapped and killed by the hunters. Thus, Aśoka's decree simultaneously took care of plants as well as animals.

Tulsī (*ocimum sanctum*) is perhaps the most sacred plant in Hindu households. It is like a small shrub from 30–60 cm in height. *Tulsī* is worshiped by the Hindus as goddess, just like

[13] *Atharvaveda*, 12: 1: 45
[14] *Atharvaveda*, 12: 1: 17 and 27
[15] *Atharvaveda*, 12: 1: 63
[16] Gosling, 2001, taken from Van Horn, 2006.
[17] *Mahābhārata*, *Anuśāsana Parva*, 10.
[18] *Padma-Purāṇa*, *Sṛṣṭi-Kāṇḍa*, Chapter 28: 19–22.
[19] Smith, 1964, p. 187.

Lakṣmī, Sītā, or Rādhā. It is common for the Hindus to worship a *tulsī* plant in the morning and feed it water while facing the Sun. When the Hindus die, people put *Tulsī* leaves, mixed with Ganges (Gaṅgā) water, in the mouth of the dead body for purification. In most *pujas* (Hindu prayer), a nectar made from milk, honey, curd, and *tulsī*, called *pancāmrt* (meaning, nectar made from five substances), is given to all participants. In a Hindu wedding, the groom, his family members, and family friends generally go to bride's house in a procession. In the midst of this joyous occasion, they first go to a *pipal* or banyan tree and pray. Similarly, the bride and groom visit a tree as a couple and pray before they enter in their house after marriage. The sacred areas of most Hindu temples are surrounded by trees.

The Hindus enforce sanctity for plants and animals by associating them with gods and goddesses. Swans are associated with Sarasvatī, mice with Lord Ganesha, snakes and bulls with Lord Śiva, lions with the Goddess Durgā, Lord Kṛṣṇa with snakes and cows, and monkeys with Lord Rāma. Cows are worshiped, so are snakes, and many other life-forms. The following is a list of some trees and their corresponding deities: Goddess Lakṣmī in *tulsī* (*Ocinlum sanctum*), Goddess Śītalā in *neem*, God Viṣṇu in *pīpal* (*Ficus religiosa*), and Lord Śiva in *Vata* (*Ficus indica*).[20] Similarly, in other interpretations, *tulasī* is beloved of Lord Kṛṣṇa; *pīpal* is inhabited by Brahmā, God Viṣṇu, and Lord Śiva; *aśoka* (*Saraca indica*) is dedicated to God Kāma; *palāś*, a tree, to the Moon; *bakula* (*Mimusops elangi*) to Lord Kṛṣṇa; and *rudrākṣa* (Eleaecarpus ganitrus) to Lord Śiva. "The Hindu spiritual heritage can provide new ways of valuing, thinking, and acting that are needed if respect for the nature is to be achieved and future ecological disasters are to be averted."[21]

Feeding weak and old animals is considered a charitable act. Like the household dogs and cats in the Western world, all animals were bestowed a proper treatment by the Hindus.[22] When Hindus eat food, they separate some food from their plate for birds, dogs, and cows. The first *chapāti* usually goes to a cow and the last one to a dog in many Hindu households.[23]

The sanctity of cows in Hinduism is well known and much publicized in the West. In legends and popular stories, the Earth is said to assume the form of a cow, especially in times of distress, to implore the help of the gods. The sacredness of the cow represents an anomaly to social scientists who wonder why in a country with hungry population the cow is virtually untouched.[24] The *Mahābhārata* tells us that the killing of a cow is wrong and has its bad consequences and causes a person to go to hell.[25] The relationship between the cow and man is not competitive, but symbiotic. The cow augments the needs of human beings by providing them

[20]Dwivedi, 1997.

[21]Dwivedi, 1997

[22]Narayanan, 2001.

[23]Most Hindu families do not keep dogs as pets. Usually, each street has a few stray dogs that are patronized by the families living on that street. These dogs recognize people who live there and protect the street from strangers at night. Mahatama Gandhi used to say that the kindness of a culture can be judged by the way they treat animals.

[24]Weizman, 1974.

[25] "All that kill, eat, and permit the slaughter of cows rot in hell for as many years as there are hairs on the body of the cow so slain." (*Mahābhārata*, 13: 74: 4.)

milk, bullocks, dung, and hides. Judging the value of Indian cattle by the western standards is inappropriate.[26] Due to inherent differences in the value systems, such an evaluation generally leads to confusion or misleading conclusions.

Kauṭilaya's *Arthaśāstra* instructed the director of forests, superintendent of cattles, horses, and elephants to prevent cruelty to animals and protection of wildlife. Punishments were inflicted on those who violated these rules. For example, killing an elephant was punished with death.[27]. Similarly, killing or injuring animals in reserve parks and sanctuary, especially those that were protected species, were punished:[28] "Elephants, horses, or animals having the form of a man, bull or an ass living in oceans, as well as fish in tanks, lakes, channels and rivers; and such game birds . . . shall be protected from all kinds of molestation. Those who violate the above rule shall be punished first with amercement. Animals useful for riding, milk, hair, or to stud were protected and hurting these animals was a crime.[29]

Because the earth is a closed system, we need processes that will sustain us in the long run. We need to find effective methods to deal with the population growth. We also must develop new technologies for food production, new family planning practices, and we must change our dietary choices. Christian J. Peters, a scientist from Tufts University, Boston, along with a team of several researchers from other universities, studies the environmental impact and food security of various diets. The purpose of this study is to compare the per capita land requirements and potential carrying capacity of the land base of the continental United States (U.S.) under a diverse set of dietary scenarios. This team considered ten diet scenarios using a biophysical simulation model. Eight of the diets complied with the 2010 Dietary Guidelines for Americans. Meat-eating diets turned out to be not good. The highest capacity turned out to be with lacto-vegetarian diet, even better than vegan diet.[30] This is basically the diet of most Hindus.

The ancient Hindus, in their analysis, observed a balance in *prakṛti* (primal nature); every species plays an important role in the balance. This balance of *prakṛti* was based on the knowledge gained through empiricism, intuition, and experimentation. Once a common connection between plants, animals, and humans was established, the coexistence of all living forms followed. Hanumāna, and Gaṇapati (Gaṇeśa), as well as trees, are revered widely by the Hindus.[31]

7.2.1 VEGETARIANISM

Livestock production accounts for 70% of all agricultural land use and 30% of the land surface of the planet, excluding the polar ice-caps. The livestock business contributes about 4.6 to 7.1

[26] Harris, The Myth of the Sacred Cow, in Leeds and Vayda, 1965.

[27] *Arthaśāstra* 2: 2: 9; Shamasastry 50, p. 49.

[28] *Arthaśāstra*, 2: 26: 1; Shamasastry, 112, p. 135.

[29] *Arthaśāstra*,4: 13: 21; Shamasastry, 235, p. 263.

[30] Peters *et al.*, 2016

[31] Mythological stories are associated with these gods. It may be demeaning to some that the Hindus gods have some animal attributes. However, most Hindus do not even think about that. Crowds in temples all over the India on Tuesday testify their reverence to Lord Hanumāna. Similarly, it is common for people to worship Lord Gaṇeśa first on any auspicious occasion. For more information, see Nelson, Lance [Editor], 1998; and Chapple and Tucker [Editors], 2000; Roy, 2005.

billion tons of greenhouse gases each year to the atmosphere, which accounts for 15% to 24% of the total current greenhouse gas production, according to a report from the United Nations.[32] The greenhouse gases released from livestock production is higher than from all forms of transportation combined. Yes, the climate change that the world is facing is directly connected to our dinner plate.

To produce one pound of beef, a steer must eat sixteen pounds of grain and soy, the remaining fifteen pounds are used to produce energy for the steer to live.[33] The production of one kilogram of beef in a US feedlot causes the emission of about 14.8 kg of CO_2. In comparison, 1 gallon of gasoline emits about 2.4 kg of CO_2.[34] An average American's meat intake is about 124 kg per year, the highest in the world. In comparison, the average for a global citizen is 31 kg a year.[35]

The *Manu-Smṛti* tells us that people who give consent to killing, who dismember a living entity, who actually kill, who purchase or sell meat, who purify it, who serve it, and who eat the meat are all sinners.[36] Bhīṣma explains to Yudhiṣṭhira that the meat of animals is like the flesh of one's own son, and a person who eats meat is considered a violent human being.[37] The *Mahābhārata*[38] also suggests that "dharma exists for the general welfare of all living beings; hence by which the welfare of all living creature is sustained, that is real dharma."

In the Hindi language, the word *ācar-vicār* (*ācar* = conduct, *vicār* = thought) signifies the connection of mind and physical body. This term is commonly used to describe the nature of a person. Diet and the state of mind are closely related. The *Vaiśeṣika-Sūtra* suggests that "improper diet instigate violence."[39]

The ancient Hindus, in their astute observation of *prakṛti* (nature) noticed a distinct role of each life-form and the importance of the "balance" in *prakṛti* was realized: "A man who does no violence to anything obtains, effortlessly, what he thinks about, what he does, and what he takes delight in. You can never get meat without violence to creatures with the breath of life, and the killing of creatures with the breath of life does not get you to heaven; therefore, you should not eat meat. Anyone who looks carefully at the source of meat, and at the tying up and slaughter of embodied creatures, should turn back from eating any meat, " suggests the *Manu-Smṛti*.[40]

Aśoka (reigned 272–232 BCE), grandson of Chandragupta (Candragupta, reigned 324–300 BCE), tried to sanctify animal life as one of his cardinal doctrines. The Greek and Aramaic edicts in Kandhar in Afghanistan show Aśoka's resolute resolve against animal killing. "The King abstains from the slaughter of living beings, and other people including the king's hunters

[32] Steinfeld, *et al.*, 2006.
[33] Kaza, 2005.
[34] Subak, 1999 and Fiala, 2008.
[35] Fiala, 2008.
[36] *Manu-Smṛti*, 5: 51–52.
[37] *Mahābhārata, Anuśāsana Parva*, 114: 11.
[38] Śānti Parva, 109: 10
[39] *Vaiśeṣika-Sūtra*, 6: 1: 7.
[40] *Manu-Smṛti*, 5: 47–49.

and fishermen have given up hunting. And those who could not control themselves have now ceased not to control themselves as far as they could. . ."[41] Aśoka issued the following decree as known in Rampurva text: "Those she-goats, ewes (adult female sheep) and sows (adult female pig), which are either pregnant or milch, are not to be slaughtered, nor their young ones which are less than six months old. Cocks are not to be caponed. Husks containing living beings should not be burnt. Forests must not be burnt either uselessly or in order to destroy living beings."[42] These were the moral codes, not only the administrative one for ruling. Kinship with nature became a basis of their kinship with God.

Hindu nonviolence made such an impact on the Muslim King Jahāngīr (r. 1605–1627 CE) that, in 1618 CE, he took a vow of nonviolence. In 1622 CE, this vow was broken when he had to pick up his gun to save his own life and his reign against his own son Khurram. This was the period when rulers used cruelty against animals and humans to demonstrate their imperial authority. Jahāngīr took this vow after killing about 17,167 life forms in 37 years. He ordered Thursday and Sunday to be holidays against the slaughter of animals and animal eating—Sunday being the birthday of his father Akbar and Thursday as the day of his accession to the throne.[43]

Obviously, the ancient Hindus were not concerned with energy or water consumption issues when they propagated vegetarianism. However, they were concerned with long life, good health for themselves, morality, animal rights, and sustaining ecology. They studied nature and came up with relationships between various life forms: plants, animals, and humans. These scientific studies led them to define a life style that led them to vegetarianism. Their concerns have a newly found meaning in today's world. Their life style, if adopted today, can easily take care of the energy crisis that the world is facing today, improve the quality of water, completely erase the problem of hunger for a while with the current population growth, and conserve ecology and nature.

7.3 LIFE IN PLANTS: SIMILARITIES WITH HUMANS

The *Manu-Smṛti* suggests that plants are susceptible to pain and pleasure. "There are various kinds of shrubs and bushy plants, and various kinds of weeds and grass, creepers and trailing plants, some of which grow from seeds and others from graft. Variously enshrouded by the quality of *tamas* (ignorance), the effects of their own acts, they retain their consciousness inward, susceptible to pleasure and pain."[44]

The *Mahābhārata* mentions a discourse between two monks, Bhṛgu and Bhārdvāja, on the epistemology of plants. Bhṛgu asks: "Sir, all plant-life and animals are guided by the same five primal elements (*pañcabhūta*). But I don't see it in the plants. Plants do not possess body heat, do not move their parts, remain at one place, and do not seem to have the five elements. Plants

[41] Sircar, 1957, p. 45.
[42] Sircar, 1957, p. 73–74. For more information, please also see, Smith, 1964; Barua, 1946.
[43] Findley, 1987.
[44] *Manu-Smṛti*, 1: 48–49.

neither hear, nor see, nor smell, nor taste. They cannot feel the touch of others. Why then you call them a component of five elements? They do not seem to possess any liquid material in them, heat of body, any earth, any wind, and any empty space. How then can plants be considered a product of five elements?"[45]

To such questions, Bhārdvāja replied: "Though trees are fixed and solid, they possess space within them. The blooming of the fruits and flowers regularly take place in them. Body heat of plants is responsible for the dropping of flower, fruit, bark, leaves from them. They sicken and dry up. This proves the sense of touch in plants. It has been seen that sound of fast wind, fire, lightening affect plants. It indicates that plants must have a sense of hearing. Climbers twine the tree from all sides and grows to the top of it. How can one proceed ahead unless it has sight? Plants therefore must have the vision. Diseased plants may be cured by specific fumigation. This proves that plants possess sense of smell and breathe. Trees drink water from their roots. They also catch diseases from contaminated water. These diseases are cured by quality water. This shows that they have a perception of taste. As we suck water through a tube, so the plants take water through their roots under the action of air in the atmosphere. Trees when cut produce new shoots, they are favored or troubled by certain factors. So they are sensitive and living. Trees take in water through roots. Air and heat combine with water to form various materials. Digesting of the food allows them to grow whereas some food is also stored."[46]

A close analysis of the above lengthy discourse between Bhrgu and Bhārdvāja depicts that the Hindus believed of life in plants, metabolism in plants, a need of food in plants, diseases in plants, and possible cures of diseases in plants using medicine. All these properties are also peculiar to animals and humans. The above quotation also tells us the Hindus' belief that plants are sensitive to sound, touch, and quality of irrigated-water just like humans.

The *Ṛgveda* mentions the hearing qualities in plants: "All plants that hear this speech, and those that have departed away, Come all assembled and confer your healing power upon this herb."[47] Thus, though plants do not have organs like humans, they do have similar functions—they breathe, eat, sleep, listen, reproduce, and die similar to humans.

The *Bṛhadāraṇyaka-Upaniṣad* compares a man to a tree: "A man is indeed like a mighty tree; his hairs are like his leaves and his skin is its outer bark. The blood flows from the skin [of a man], so does the sap from the skin [of a tree]. Thus, blood flows from a wounded man in the same manner as sap from the tree that is struck. His flesh [corresponds to what is] within the inner bark, his nerves are as though the inner fibers [of a tree]. His bones lie behind his flesh as the wood lies behind the soft tissues. Marrow [of a human bone] resembles with the pith [of a tree]."[48]

The *Chāndogya-Upaniṣad* recognized the presence of life in plants. "Of this great tree, my dear, if someone should strike at the root, it would bleed, but still live. If someone should

[45] *Mahābhārata*, *Śānti-Parva*, 184: 6–18.
[46] *Mahābhārata*, *Śānti-Parva*, 184: 10–18.
[47] *Ṛgveda*, 10: 97: 21
[48] *Bṛhadāraṇyaka-Upaniṣad*, 3: 9: 28.

strike its middle, it would bleed, but still live. If someone should strike at its top, it would bleed, but still live. Being pervaded in *ātman* [soul or self], it continues to stand, eagerly drinking in moisture and rejoicing. If the life leaves the one branch of it, then it dries up. It leaves a second; it dries up . . . Verily, indeed, when life has left it, this body dies."[49]

The presence of life in plants and similarities between plants and humans are quite clearly defined in the ancient Hindu literature. It was only natural for someone to test these ideas. A major attempt came from Sir J. C. Bose (1858–1937 CE) of Calcutta during the earlier parts of the twentieth century. Though Bose was trained in physics, he ventured into the discipline of botany and made valuable contributions. As a result of his contributions, Bose was knighted by the British monarchy in 1917 and was elected a Fellow of the Royal Society of London in 1920. He was thus the third Indian and the first natural scientist from India to be so honored. The first Indian to get elected as a Fellow of the Royal Society was Engineer, Adaseer Cursetjee, in 1841, and the second was mathematician Srinivas Ramanujan, elected in 1918. Bose demonstrated life essences in plants with his experiments during the late nineteenth century. His contributions are: [50]

1. *Power of response* (reflex action) In a mimosa plant, if a single leaf is touched, all the leaves in the branch slowly close. Bose devised an instrument to study such a behavior in other plants. He concluded that all plants, like human beings, possess reflex actions.

2. *Food Habits* Bose demonstrated that plants need food, like humans; nourishment of a plant comes from its roots as well as from its body. Bose experimentally demonstrated that, like humans, plants nourish themselves from a distribution of nutrients through a continuous cycle of contraction and expansion of cells. If a poison is administered to a plant, it may succumb to it. On the removal of poison, it may gradually revive and become normal. Plants also sway abnormally like a drunk person if treated with an alcoholic substance and become normal when the cause is removed.

3. *Mind Activity* A sun-flower plant adjusts itself throughout the day so that it always faces the Sun. Many other plants close their leaves during the night and turn to face a brighter light source if the light is not uniformly distributed, indicating a mind activity in plants.

4. *Nerves* Plants also have nerves and feel sensation through leaves, branches, or trunks. If certain areas of a plant are made numb with ice, less pain is felt in the plant.

Scientific American, a popular science magazine in America, published an article on J.C. Bose and summarized his work in the following words: "By a remarkable series of experiments conducted with instruments of unimaginable delicacy, the Indian scientist [J. C. Bose] has discovered that plants have a nervous system . . . With the possibilities of Dr. Bose's crescograph,

[49] *Chāndogya-Upaniṣad*, 6: 12: 1–3.

[50] for more information, read Subrata Dasgupta, 1998.

in less than quarter of an hour the action of fertilizers, foods, electric currents and various stimulants can be fully determined."[51]

[51] *Scientific American*, April 1915.

CHAPTER 8

Medicine

Diseases are as old as life itself. Manu (progenitor of humanity), or Adam and Eve, must have also needed the science of medicine. The problems of colds, fever, headache, and exhaustion has been experienced by people at one time or another. The ancient Hindus developed a system that is popularly known as Ayurveda, literally translated as the science of life. According to Suśruta, a noted ancient Hindu surgeon, Brahmā (creator of the universe) was the first to inculcate the principles of Ayurveda and taught it to Prajāpati. The Aśvin Kumaras, two brothers, learned from Prajāpati and taught it to Indra. Indra taught it to Dhanvantari who later taught it to Suśruta.[1] Dhanavantari's status in India is similar to Aesculapius in the Western world. Caraka, another ancient Hindu physician, considered Ayurveda to be eternal:[2] "The science of life has always been in existence, and there have always been people who understood it in their own way: it is only with reference to its first systematized comprehension or instruction that it may be said to have a beginning."

The *Ṛgveda* suggests the human life expectancy to be 100 years.[3] The *Yajurveda-Śukla* tells everyone to aspire to have an active life till they reach 100.[4] The *Upaniṣads* also suggest us that we should expect to live an active life for hundred years. "Even while doing deeds here, one may desire to live a hundred years," suggests the *Iśā-Upaniṣad*.[5] During the ancient period, in most cultures, the life expectancy was about 40 years of age. However, among the ancient Hindus, it was the second stage of life at that age. Marco Polo, when he visited India, has mentioned of the long life of the Hindus.

The *materia medica* of the Indus-Sarasvatī region is particularly rich due to a large number of rain forests around the country. The Indian climate is unique, as all four seasons are experienced. The abundance of *materia medica* has been recorded by several foreigners who visited India during the ancient and medieval periods. Although India's total land area is only 2.4% of the total geographical area of the world, today it accounts for 8% of the total global biodiversity with an estimated 49,000 species of plants of which 4,900 are endemic, implying their presence only in India or nearby regions.[6] There are around 25,000 effective plant-based formulations available

[1] *Suśruta-Saṁhitā, Sūtrasthānam*, 1: 16.

[2] *Caraka-Saṁhitā, Sūtrasthānam*, 30, 26–28.

[3] "That Indra for a hundred years may lead him safe to the farther shore of all misfortune." *Ṛgveda*, 10: 161: 3. "With the most saving medicines which thou givest, Rūdra, may I attain a hundred winters." The *Ṛgveda*, 2: 33: 2. As a comparison, the life expectancy among the ancient Egyptian was only about mid-thirties while for the Roman it was mid-forties.

[4] *Yajurveda-Śukla*, 40: 2.

[5] *Iśā-Upaniṣad*, 2.

[6] Ramakrishnappa, K., 2002. This report can be accessed at http://www.fao.org/DOCREP/005/AA021E/AA021e00.htm.

in Indian medicine. It is estimated that there are over 7,800 medicinal drug manufacturing units in India, which consume about 2,000 tonnes of herbs annually.[7] The *Caraka-Saṁhitā* lists 341 plants and plant products for use in medicine. Suśruta described 395 medicinal plants, 57 drugs of animal origin and 64 minerals and metals as therapeutic agents.[8]

The ancient Hindus understood that disease is a process that evolves over time. They tried to prevent diseases by using proper hygiene, a healthy lifestyle, a balanced diet, yoga, internal and external medicine, surgery, and prayer. Good health was considered as essential to achieve *mokṣa*.[9] Detailed experiments were performed to evolve new medical knowledge. For example, Suśruta selected a cadaver, cleaned it, and wastes were removed. Afterward, the body was wrapped in grass and placed in a cage to protect it from animals. The cage was then lowered into a river or a rushing stream of water. This was done to ensure constant cleaning. Once the decay process took a toll on the body, it was taken out of water. Skin, fat, and muscles were brushed away from the corpse with grass brushes, and the human anatomy was observed.[10]

In evaluating the diagnostic techniques of the ancient Hindus, Erwin Heinz Ackerknecht, in his book *A Short History of Medicine*, writes that "[d]iagnostic were highly developed. The Indian healer used questioning, very thorough inspection (e.g., to diagnose pulmonary consumption from the loss of weight), touch (including observation of the pulse), and examination with other senses, such as tasting the urine for diabetes. Indians knew the sweet taste of diabetic urine long before Europeans did."[11]

8.1 DOCTORS, NURSES, PHARMACIES, AND HOSPITALS

During the Vedic period, doctors were divided into surgeons (*śalya-vaidya*) and physicians (*bhiṣak*).[12] These doctors were well respected in the society. "He who stores herbs [medicine man] is like a king amid a crowd of men, Physician is that sage's name, fiend-slayer, chaser of disease," suggests the *Ṛgveda*.[13]

The inscriptions of King Aśoka (third century BCE) refer to the cultivation of medicinal plants and the construction of hospitals in his kingdom. The rock edict of Girinar in India, erected by Aśoka (r. 269–232 BCE), indicates that separate hospitals for humans and animals were built. Aśoka suggests *ahiṁsā* in the treatment of all animals: "Meritorious is abstention from the slaughter of living beings."[14] Indeed, hospitals were build all over India from the earliest periods.

[7] Mukherjee and Wahile, 2006.

[8] Mukherjee and Wahile, 2006. Krishnamurthy has given quite a different number for the medicinal plants. According to Krishnamurty, Suśruta defined 793 plants for medical usages. Krishnamurthy (1991) has compiled the list of these plants in the ninth chapter of his book. Krishnamurthy, 1991.

[9] *Caraka-Saṁhitā*, *Sūtrasthānam*, 1: 15–16.

[10] Rajgopal *et al.*, 2002.

[11] Ackerknecht, 1982, p. 38.

[12] Mukhopādhyāya, 1994.

[13] *Ṛgveda*, 10: 97: 6.

[14] Sircar, 1957, p. 47.

Caraka defines following qualities of a doctor: good knowledge of the medical texts; experienced in curing various diseases, skillful in the preparation of drugs, and their uses; good in quick thinking in difficult situations; and good in hygiene.[15] The doctor should also have control over his hands when in surgery or difficult situations. A person with shaking hands is not a good surgeon.[16] "Drugs are like nectar; administered by the ignorant, however, they become weapons, thunderbolt or poison. One should therefore shun the ignorant physician," suggests Suśruta.[17] Emphasizing constant practice in diagnosis, Suśruta suggests that "the physician who studies the science of medicine from the lips of his preceptor, and practices medicine after acquired experience in his art by constant practice is the true physician."[18]

Caraka advised people to avoid incompetent doctors,[19] as it can even aggravate their condition.[20] Caraka has provided a detailed code of conduct for doctors which is similar to the Hippocratic Oath in the Western world.[21] This code of conduct forbids doctors to indulge in money-making and lust: "He, who practices not for money nor for caprice but out of compassion for living beings is the best among all doctors."[22] And, most importantly, a doctor must continue to help his patients, despite adversity. This is the only way to save the life of a person.[23] Physicians were instructed to be friendly, kind, eager to help underprivileged people, and remain calm in difficult diseases.[24] "The underlined objective of a treatment is kindness. Kindness to another human being is the highest *dharma*. A doctor becomes successful and finds happiness with this objective."[25] Doctors were advised to concentrate only on the treatment of the patient and not on materializtic rewards, the patient was advised to reward the doctor with money that he can afford and to be respectful toward the doctor.[26]

According to an ancient legend, Jivaka Kumarabhacca was a revered physician who provided medical care to Lord Buddha and his disciples. Jivaka studied medicine under the legendary and revered physician Ātreya, in the city of Taxila. After many years of training, one day Jivaka decided to have his own independent practice and shared his views with his teacher. Ātreya decided to take an exam before he could allow Jivaka to have his own practice. He asked Jivaka to go around the *āshram* (monastery) and collect all plants with no medicinal value. Jivaka walked around and toiled to look for a plant with no medicinal value. After considerable efforts and somewhat dejected, he went back to Ātreya with no plant in his hand and told him that he

[15] *Caraka-Saṁhitā*, *Sūtrasthānam*, 9: 6.
[16] *Caraka-Saṁhitā*, *Sūtrasthānam*, 29: 5–6.
[17] *Suśruta-Saṁhitā*, *Sūtrasthānam*, 3: 51.
[18] *Suśruta-Saṁhitā*, *Sūtrasthānam*, 4: 7
[19] *Caraka-Saṁhitā Sūtrasthānam*, 9: 15–17.
[20] *Caraka-Saṁhitā*, *Sūtrasthānam*, 16: 4.
[21] *Caraka-Saṁhitā*, *Vimānasthānam*, 8: 13.
[22] *Caraka-Saṁhitā*, *Vimānasthānam*, 8: 13.
[23] *Caraka-Saṁhitā*, *Sūtrasthānam*, 1: 134.
[24] *Caraka-Saṁhitā*, *Sūtrasthānam*, 9: 26.
[25] *Caraka-Saṁhitā*, *Cikitsāsthānam*, 4: 63.
[26] *Caraka-Saṁhitā*, *Cikitsāsthānam*, 4, 55.

could not find even a single plant with no medicinal value. Ātreya was pleased with the answer and gave his blessing to Jivaka to start his own practice for the welfare of human beings.

Caraka and Suśruta emphasized that a good doctor must have all needed equipment with him.[27] Similarly, Suśruta suggested multitude of surgical tools, leeches, fire for sanitizing tools, cotton pads, suture threads, honey, butter, oil, milk, ointments, hot and cold water, etc. for a surgeon. For the tools, roundhead-knife, scalpel, nail-cutter, finger-knife, lancet that is as sharp as the leaf of the lotus, single-edge knife, needle, bistoury, scissors of various shapes, curved bistoury, ax-shaped hammer, trocar, hooks, narrow-blade knife, toothpick, etc. Surgeon must know to hold these instruments properly for surgery[28] and surgery should be avoided when one could effectively use leeches to do the same job.[29]

"One who knows the text but is poor in practice gets confounded on seeing a patient; he is like a coward on the battle field. One who is bold and dexterous but lacking in textual knowledge fails to win the approval of peers and risks capital punishment from the king. Having half-knowledge, implying either textual or practical knowledge, is a physician who is unfit for the job and is like a one-winged bird," suggests Suśruta.[30] In the views of Suśruta, a person with just theoretical knowledge in ayurveda only, is "like an ass that carries a load of sandal wood [without ever being able to enjoy its pleasing scent]."[31]

Suśruta advised that doctors must involve all five senses in the physical examination of the patient: inspection, palpation, auscultation (listening to the sound of the heart, lungs, etc.), taste, and smell.[32] For incisions, surgeons were advised to use the tool rapidly only once to desirable depth.[33] Obviously, this was done to avoid or minimize pain to patients. It was also advised that the surgeon must be courageous, must be quick in action and decision-making, self confident, should not sweat in tense situations, his hands should not shake, and he should have sharp instruments.[34]

Caraka advised doctors to differentiate between the diseases that can be cured and those that cannot be cured.[35] He advised doctors to not treat patients with incurable diseases in advanced stages and alleviate the suffering of the patient.[36] Since disease is a process with various stages, he suggested that even serious diseases can be cured if caught in their early stages. Proper medication at the early stage and proper lifestyle is essential to control these serious ailments.[37]

[27] *Caraka-Saṁhitā*, *Sūtrasthānam*, 16: 1; *Suśruta-Saṁhitā*, *Sūtrasthānam*, 5: 6.

[28] *Suśruta-Saṁhitā*, *Sūtrasthānam*, 8: 3–4.

[29] *Suśruta-Saṁhitā*, *Sūtrasthānam*, 13: 3–4.

[30] *Suśruta-Saṁhitā*, *Sūtrasthānam*, 3: 48–50

[31] *Suśruta-Saṁhitā*, *Sūtrasthānam*, 4: 2

[32] *Suśruta-Saṁhitā*, *Sūtrasthānam*, 10: 4.

[33] *Suśruta-Saṁhitā*, *Sūtrasthānam*, 5: 7.

[34] *Suśruta-Saṁhitā*, *Sūtrasthānam*, 5: 5–6.

[35] *Caraka-Saṁhitā*, *Sūtrasthānam*, 18: 39.

[36] *Caraka-Saṁhitā*, *Sūtrasthānam*, 18: 40–44. This is a sensitive issue that has moral, cultural, and moral implications. Most people incur major medical expenses in the last five years of their lives. Should we treat a patient with no or little chance of recovery? The *Edwin Smith Papyrus* has dealt with the same issue more than 3000 years ago in Egypt. This issue is still under discussion in many countries around the world.

[37] *Caraka-Saṁhitā*, *Sūtrasthānam*, 18: 38.

It is essential for a doctor to recognize the disease by knowing the symptoms of the ailment. However, since the number of diseases is very large, a doctor should not feel ashamed if he cannot identify the disease from the symptoms.[38] This is an example of honesty that is expected of all physicians.

Doctors were responsible for treatment and were punished for wrongdoings. The penalty for such wrongdoings depended on the extent of damage. Humans received more attention than animals in their medical treatment.[39] Physicians were not above the law; their greed and carelessness were checked with laws during the period of Chandragupta Maurya (Candragupta, reigned 322–298 BCE) in India.[40]

According to Caraka, the following are the four constituents that are crucial for the quick recovery of a patient: a qualified doctor, a quality drug, a qualified nurse, and a willing patient.[41] This clearly signifies the importance of nursing in medical care. The roles of a nurse are: the complete knowledge of nursing, skillful, affection for the patient, and cleanliness.[42] A good patient is defined as someone with a good memory, follows the instructions of the doctor, free of fear, and can explain the symptoms of the disease well. This combination of doctor, nurse, drug, and patient is essential for a successful cure. Otherwise, the cure is incomplete.[43]

Patients were advised to ignore doctors who practice just for money, were not qualified, or were too proud of their status.[44] Patients should go to a doctor who has studied *śāstra* (texts), is astute, is pure, understands the job well, has a good reputation, and has good control of his mind and body, suggested Caraka.[45] A good doctor should be respected like a guru.[46] "The patient, who may mistrust his own parents, sons, and relatives, should repose an implicit faith in his own physician, and put his own life into his hands without the least apprehension of danger; hence a physician must protect his patient as his own begotten child," suggest Suśruta.[47]

Pharmacies prepared medicines as powders, pastes, infusions, pills, confections, or liquids using chemical techniques of oxidation, fermentation, and distillation. Pharmacists were trained in grafting, general care of plants, extract juices from flowers, concocting various medical preparations, and making medicinal *bhasms* and extracts.

In providing the design of a hospital, Caraka suggests that it should have good ventilation and the rooms should be big enough for patients to walk comfortably. The rooms should not experience the intense glare of sunlight, water, smoke, and or dust. The patient should not experience any unpleasant sound, touch, view, taste, or smell. Expert chefs, masseuses, servants,

[38] *Caraka-Saṁhitā*, *Sūtrasthānam*, 18: 45–46.
[39] *Manu-Smṛti*, 9: 284.
[40] Kauṭilaya's *Arthaśāstra*, 144.
[41] *Caraka-Saṁhitā*, *Sūtrasthānam*, 9: 3.
[42] *Caraka-Saṁhitā*, *Sūtrasthānam*, 9; 8.
[43] *Caraka-Saṁhitā*, *Sūtrasthānam*, 9: 9–13.
[44] *Caraka-Saṁhitā*, *Sūtrasthānam*, 29: 12.
[45] *Caraka-Saṁhitā*, *Sūtrasthānam*, 29: 13.
[46] *Caraka-Saṁhitā*, *Sūtrasthānam*, 4: 51.
[47] *Suśruta-Saṁhitā*, *Sūtrasthānam*, 25: 42–45.

nurses, pharmacists, and doctors should be hired to take care of the patients. Birds, deer, cows and other animals, along with singers and musicians should be kept in the hospital compound.[48]

Hospitals should have enough quality food in the storage room, stored drugs, and a compound with drug plants. The doctors must be intelligent in knowing the books as well as equipment. Only then, they were allowed to help a patient.[49]

8.2 AYURVEDA

Ayurveda is the Hindu science of healing and rejuvenation which is curative as well as preventive. It is a holistic medicine that focuses on mind and body and a person needs to practice the doctrines of ayurveda when healthy to avoid sickness. Dietary measures and lifestyle changes are recommended to delay the aging process at the cellular level and to improve the functional efficiency of body and mind. The ancient Hindus recognized the self-healing capacity of our body and focused on creating a balance that allows the body to cure itself.

Ayurveda is a combination of two words–*āyur* and *veda*. *Āyur* means the span of life while *veda* means unimpeachable knowledge or wisdom. Ayurveda signifies life science for the well being of body and mind. This stemmed out of Hindus' belief that health and disease co-inhere in body and mind. Ayurveda has its origin in the *Vedas* for the "principal elements of its general doctrines."[50] In the Hindu tradition, Ayurveda is considered as an *Upa-Veda* (subordinate *Veda*) that deals with medicine and medical treatment for good health, longevity, and elimination of disease. It is considered to be in well-evolved form at the time of *Atharvaveda*. Suśruta wrote: "The Ayurveda originally formed one of the subsections of the *Atharvaveda* and originally consisted of 100,000 verses."[51]

Ayurveda uses herbal medicine, dietetics, surgery, psychology, and spirituality to cure diseases. *Pañca-karma* (detoxification of body), laxatives, herbal oil massages, and nasal therapies are some of the treatments. Though ancient in origin, it is still the most popular system of medicine in modern India. Several Indian universities provide doctoral degree programs in Ayurveda. These universities train doctors who practice throughout India and provide low cost medical care to poor people. These days, Banaras, Chennei, Haridwar, Lucknow, Patna, and Trivandrum are some of the prominent cities that house universities to train ayurvedic doctors.

Caraka suggested that physical ailments should be cured by proper diet, changing daily routines that included moderate physical exercise,[52] and medicine, while the mental diseases should be cured by self control, meditation, and even by charms (*bhuta-vidyā*).[53] Caraka defines

[48] *Caraka-Saṁhitā, Sūtrasthānam*, 15: 7.
[49] *Caraka-Saṁhitā, Sūtrasthānam*, 16: 1.
[50] Filliozat, 1964, p. 188.
[51] *Suśruta-Saṁhitā, Sūtrasthānam*, 1: 3.
[52] *Caraka-Saṁhitā, Sūtrasthānam*, 7: 30.
[53] *Caraka-Saṁhitā, Sūtrasthānam*, 1–8.

that the purpose of Ayurveda is to preserve the well-being of a healthy person and to get rid of the diseases of a sick person.[54]

In the ayurvedic tradition, the human body is considered as a conglomeration of five elements, known as *mahābhūtas*—earth (*pṛthivī*), air (*vāyu*), water (*apaḥ*), fire (*tejas*), and space (*ākāśa*). The five *mahābhūtas* produce seven *dhātus*: *rasa* (juice-plasma), *rakta* (blood), *māṃsa* (flesh), *medas* (fat), *asthi* (bone), *majjā* (marrow), and *śukra* (reproductive fluids in male and females).[55] The five *mahābhūta*, or the seven *dhātu*, constitute the *prakṛti* (attributes) of a person. These *dhātu* are a product of the five elements. When in equilibrium, these *dhātus* make a person happy and healthy. The various combinations of these *dhātu* constitute the tri-*guṇas* (three-attributes) in a person: *sattva*, *rājas*, *tāmas*. A person with dominant *sattva* values truth, and honesty; *rājas* value power, prestige, and authority. A *tāmas* dominant person lives in ignorance, fear, and servility.

The basic theory of Ayurveda rests on the three humors (*tridoṣa*; *tri* = three, *doṣa* = the literal meaning is defect, popularly known as elements)–*Kapha* (phlegm), a heat regulating system, and mucous and glandular secretion; *vata* (wind), a nerve force; and *pitta* (bile), metabolism and heat production. These three *doṣa* control the normal functions of a human body and a balance is necessary. The term *doṣa* is generally translated as "corrupting agents," "defect," or "imperfection." However, in Ayurveda, these *doṣa* are the keys to good health. The theory of *doṣa* is pretty complex as it "is affected by an almost infinite number of exogenous factors and combinations of factors that make up human ecology: diet, rate, time, and context of food consumption, climate, direction of wind, age, behavioral patterns, rest, accidents, exercise, state of mind, and so forth."[56] An imbalance of these three *doṣas* causes most diseases. It is the purpose of all treatments to keep the balance of the *doṣas* by balancing the seven *dhātus*.[57] The *Suśruta-Saṁhitā* gives various symptoms of the deficiency of these *tridoṣas* and their effects on body.[58]

Balance is essential in the natural world and is a key concept in Hindu philosophy. It is the balance of body and mind in *āsana* that is crucial in yoga. It is the balance of *tridoṣas* in ayurveda that is crucial in the well-being of a person, just like the *yin* and *yang* in Chinese philosophy. Therefore, before ayurvedic doctors look for the cure of a disease, they inquire about the patient, his/her constitution, eating habits, and nature to know the imbalance in *dhātu*. For example, in describing the cause of diabetes and its cure, Suśruta and Caraka attribute lack of exercise, laziness, sweet foods, alcoholic foods and beverages, elevation of *kapha*, and excess of newly harvested food grains as the cause. A continual balance of *tridoṣa* results in contentment, longevity, and enlightenment. We eat food to nourish the seven *dhātus* in our body. By changing diet, people can change the equilibrium of *dhātus* and thus their attributes. Therefore, among the Hindus, it is a customary to describe the attributes of a person from a combination of two

[54] *Caraka-Saṁhitā*, *Sūtrasthānam*, 30: 26.

[55] For the roles of these *dhātu*, read *Suśruta-Saṁhitā*, *Sūtrasthānam*, Chapter 15.

[56] Alter, 1999; Lad, 1984; and Zysk, 1991.

[57] *Caraka-Saṁhitā*, *Sūtrasthānam*, 16: 35–36.

[58] *Suśruta-Saṁhitā*, *Sūtrasthānam*, Chapter 15.

words, *ācar-vicār*, describing the thinking and eating habits of a person. Suśruta suggested eating in moderation for a healthy life.[59] Caraka clearly regulates the quantity of food that a person should eat: "The food must be digested in time and should not cause any inconvenience in activity of the person."[60]

Patients should seek effective drugs immediately after recognizing a sickness.[61] Thus, quick diagnosis is the key in fighting against a disease. Following are the typical drugs used in ayurveda:

1. Animal products (honey, milk, blood, urine, fat, horn, etc.)
2. Minerals and metals (salt, gypsum, alum, gold, lead, copper, iron, phosphorous, sulfur, potassium, calcium, iodine, lime, mud, precious stones, etc.)
3. Herbs and plants (for example, aloe vera, barley, beans, cinnamon, clove, coriander, cumin, black pepper, fenugreek, neem, etc.)

In Ayurveda, metals are oxidized, reduced, and transformed into a chemical compound that is non-toxic, to be used as a drug.[62] Several elements like gold, lead, copper, iron, phosphorous, sulfur, potassium, calcium, and iodine are processed with physio-chemical techniques. Water, sunlight, milk, honey, and fermented extracts of fruits are also used as medicine in Ayurveda. For example, *Drākṣaśava*, that is an extract of grape and apple-vinegar along with several minerals, with a specific fermentation technique, is perhaps the best natural medicine for indigestion. It cleans the blood, improves the functioning of kidneys and liver, improves the pH value of urine, and provides minerals. It is not only a curative medicine but could also be used on a regular basis as a preventive medicine for good health.

Minerals are also used for curative purposes; *śilājīta*, perhaps the most important one, is a gelatine substance secreted from mountain stones during the scorching heat of June and July in Northern India, and contains traces of tin, lead, copper, silver, gold, and iron. *Śilājīta* has heat-producing and body purifying properties and is used in the treatment of diabetes, leprosy, internal tumor, and jaundice.[63]

The daily routine (*dinacarayā*) is crucial for good health. Caraka deals with the issue of lifestyle in great detail in the first chapter of *Caraka-Saṁhitā*.[64] Suśruta also devoted a long chapter to prescribe a good lifestyle for a healthy person.[65] Almost as a rule, it was considered good to get up early in the morning and practice yoga and meditation after toiletry functions and bath. To optimize health and as preventive measures for various body organs and for general health, dental cleaning with herbs, mouthwash, tongue scraping, daily bath, massage, clean clothes,

[59] *Suśruta-Saṁhitā*, *Sūtrasthānam*, 46: 145.
[60] *Caraka-Saṁhitā*, *Sūtrasthānam*, 5: 4.
[61] *Caraka-Saṁhitā*, *Sūtrasthānam*, 11: 63.
[62] For more information, read Zimmer, 1948.
[63] *Suśruta-Saṁhitā*, *Cikitsāstānam*, Chapter 13.
[64] *Caraka-Saṁhitā*, *Sūtrasthānam*, sections 5, 7, and 8.
[65] *Suśruta-Saṁhitā*,*Cikitsāstānam*, 24: 3–132.

good shoes, daily exercise, and good thought processes were suggested. The role of the mind was known to the ancient Hindus. Thus, they suggested good thought process as essential to good health.

Mouth hygiene received much attention in ancient writings. Caraka suggested to brush the teeth twice a day[66] and clean the tongue using a metallic scraper.[67] He also suggests many herbs such as clove, cardamom, beetle nut, and nutmeg to keep in the mouth for a pleasant breath.[68] Suśruta provides the following instructions for the hygiene of mouth. "A man should leave his bed early in the morning and brush his teeth. The tooth-brush should be made of a fresh twig of a tree or a plant grown on a commendable tract and it should be straight, not worm-eaten, devoid of any knot or at most with one knot only (at the handle side), and should be twelve fingers in length and like the small finger in girth."[69] "The teeth should be cleansed daily with (a compound consisting of) honey, powdered *trikaṭu*, *trivarga*, *tejovatī*, *saindhave* and oil. Each tooth should be separately cleansed with the preceding cleansing paste applied on (the top of the twig) bitten into the form of a soft brush, and care should be taken not to hurt the gum any way during the rubbing process. This tends to cleanse and remove the bad smell (from the mouth) and uncleanliness (of the teeth) as well as to subdue the *kapha* (of the body). It cleanses the mouth and also produces a good relish for food and cheerfulness of mind,"suggests Suśruta.[70] Afterward, the person should cleanse the tongue by scraping it with gold, silver, or wood foil.[71]

8.2.1 *PAÑCA-KARMA*

Pañca-karma is an ayurvedic treatment to cleanse the body system to remove toxins. This treatment has three phases: *purva-karma* (preparation), *pradhā-karma* (main treatment), and *paśca-karma* or *Uttara-karma* (post-treatment).[72] Before the actual process, a dietary and herbal treatment is prescribed to cleanse the intestinal system, reduce the excess or vitiated *doṣas*, and increasing the gastric fire. In the oleation process, oily substances like sesame, flaxseed, or mustard oil or ghee are taken internally or externally in massage and sudation (sweating) techniques are used. Ghee is clarified butter by removing its water contents and milk solids. It becomes lactose-free and can be used by people with lactose intolerance. Ghee is antimicrobial and antifungal and, therefore, can be preserved without refrigeration for an extended period.[73] The oleation process lubricates bodily tissues and should be taken in moderation especially people with *kapha*. The exact prescription in *purva-karma* depends on the individual and the diet includes simple foods

[66] *Caraka-Saṁhitā, Sūtrasthānam*, 5: 71.
[67] *Caraka-Saṁhitā, Sūtrasthānam*, 5: 74.
[68] *Caraka-Saṁhitā, Sūtrasthānam*, 5: 76.
[69] *Suśruta-Saṁhitā, Cikitsāstānam*, 24: 3.
[70] *Suśruta-Saṁhitā, Cikitsāstānam*, 24: 6–7.
[71] *Suśruta-Saṁhitā, cikitsāsthānam*, 24: 11–12.
[72] Ninivaggi, 2008, p. 207
[73] Ninivaggi, 2008, p. 209.

such as *khicaṛī*, a combination of basmati rice, split mung lentils and some mild spices, such as turmeric, cumin and coriander.

The five actions of *Pañca-karma* are as follows: *vamana*, *virecana*, *vasti*, *nasya*, and *rakta-moksha*.[74] These actions are provided by Suśruta. First, the body is detoxified by inducing vomiting using herbal mixture (*vamana*), then an herbal mixture is taken for numerous bowel movements (*virecana*), oil or herbs are administered anally for some time and expelled (*vasti*). *Nasya* is practiced by using *neti-kriyā* (luke warm water passed through the sinus) and oil is applied inside to clean the sinus system. In some special cases, the blood is purified or even taken out using needles, incisions, or by leeches.[75]. In the post-treatment, dietary and lifestyle lessons are provided to have balanced *dinacaryā*.

Pañca-karma is advised for a variety of medical conditions: arthritis, rheumatism, respiratory disorders, gastrointestinal disorders, menstrual problems, obesity, etc. A team of Harvard Medical School in Massachusetts conducted a clinical study of this practice to observe the role of psychosocial factors in the process of behavior change and salutogenic process. The 20 female participants underwent in a 5-day ayurvedic cleansing retreat program. Measurements were taken before the program, immediately after the program, and after three months for the quality of life, psychosocial, and behavioral changes. This study indicates that *pañca-karma* "may be effective in assisting one's expected and reported adherence to new and healthier behavior patterns."[76]

8.3 SURGERY

Suśruta, in his *Suśruta-Saṁhitā*, describes some one hundred medical surgical instruments of metals in various shapes and sizes. These include probes, loops, hooks, scalpels, bone-nippers, scissors, needles, lancets, saws, forceps, syringes, rectal speculums, and probes. Many surgical operations including those for hernia, amputation, tumor, restoration of nose, skin grafting, extraction of cataracts, and caesarean section are described in the *Caraka-Saṁhitā* and/or the *Suśruta-Saṁhitā*. In both books, blunt as well as sharp tools depending on the needs are described. Some of these instruments are sharp enough to dissect a hair longitudinally.[77] Suśruta performed lithotomy, cesarean section, excision of tumors, and the removal of hernia through the scrotum.[78]

New doctors practiced incisions first on plants or dead animals. The veins of large leaves were punctured and lanced to master the art of surgery. Cadavers were advocated as an indispensable aid to the practice of surgery. Bandaging, amputations, and plastic surgery were first practiced on flexible objects or dead animals just as is done by medical students today. Suśruta

[74] Ninivaggi, 2008, p. 212.

[75] Ninivaggi, 2008, p. 219

[76] Conboy, Edshteyn, and Garivsaltis, 2009.

[77] For a detailed account of the tools, see Mukhopādhāya, 1994.

[78] Singh, Thakral, and Deshpande, 1970. It is an excellent article with basic information.

suggests that "a surgeon in seeking a reliable knowledge must duly prepare a dead body and carefully ascertain from his own eyes what he learned with books."[79]

Suśruta classified surgery into eight parts: extraction of solid bodies, excising, incising, probing, scarifying, suturing, puncturing, and evacuating fluid.[80] He designed his surgical instruments based on the shapes of the beak of various birds and the jaws of animals. In his attempt to explain surgery, Suśruta writes: "The scope of surgery, a branch of medical science, is to remove an ulcer and extraneous substance such as fragments of hay, particles of stone, dust iron, ore, bone; splinters, nails, hair, clotted blood or condensed pus, or to draw out of the uterus a dead fetus, or to bring about safe parturitions in cases of false presentation, and to deal with the principle and mode of using and handling surgical instruments in general, and with the application of cautery and caustics, together with the diagnosis and treatment of ulcers."[81]

All surgical procedures were defined in terms of three stages of the process: Pre-operative measures, operation, post-operation measures. This included the preparation of the surgical room, the patient, and the tools. Starvation or mild dieting was advised on the day of surgery. Intoxicating beverages were advised to avoid the pain of surgery. In the post-operative measures bandaging, redressing, healing and cosmetic restoration were also involved.[82] Anesthesia, antiseptic, and sterile techniques were used by the ancient Hindus.

Suśruta described fourteen different kinds of bandages (*bandha*).[83] On the harmful effects of non-bandaging, Suśruta suggests that flies and gnats are attracted to the wound. Foreign matters, such as dust and weeds can settle on the wound. Sweat, heat and cold can make the wound malignant.[84] Suśruta advises doctors to clean wounds first to get rid of dust, hairs, nails, loose pieces of bones, and other foreign objects before suturing to avoid suppuration. If not done, it can cause severe pain and other problems.[85] Suśruta found innovative ways to perform various surgical procedures: horse hairs were used as suture material while black carpenter ants were used in closing incisions in soft tissues. Blood or pus was drawn from body by using leeches.

The amputation of the leg and other body parts was done during the *Ṛgvedic* period. The *Ṛgveda* tells us that the lower limb of Queen Viśpalā was severed in King Khela's battle. Āsvins connected an artificial limb to her at night and she was able to fight the battle again:[86]

A neolithic 4000 year-old skeleton with a multiple-trepanated skull was recently found in Kashmir. The trepanation on this skull was perhaps accomplished using drills of various diameters and could be a result of an elaborate medico-ritual ceremonial procedure.[87] Similarly, nearly perfect tiny holes as dental treatment were found in excavations in the Mehrgarh site,

[79] *Suśruta-Saṁhitā*, *Śarīrsthānam*, 5: 6.
[80] *Suśruta-Saṁhitā*, *Sūtrasthānam*, 5: 5
[81] *Suśruta-Saṁhitā*, *Sūtrasthānam*, I: 4.
[82] *Suśruta-Saṁhitā*, *Sūtrasthānam*, 5: 3.
[83] *Suśruta-Saṁhitā*, *Sūtrasthānam*, 18: 14–18; also read Kutumbiah, 1962, p. 163.
[84] *Suśruta-Saṁhitā*, *Sūtrasthānam*, 18: 23.
[85] *Suśruta-Saṁhitā*, *Sūtrasthānam*, 25: 18.
[86] *Ṛgveda*, 1: 116: 15.
[87] Sankhyan and Weber, 2001.

near Baluchistan, that are about 7,500 to 9,000 years old. Eleven drilled molar crowns from nine adults were studied from this site. The drilled holes are about 1.3–3.2 mm in diameter and angled slightly to the occlusal plane, with a depth of about 0.5 to 3.5 mm.[88] Cavity shapes were conical, cylindrical or trapezoidal. At least in one case, "subsequent micro-tool carving of the cavity wall" was performed after the removal of the tooth structure by the drill.[89] In all cases, marginal smoothing confirms that drilling was performed on a living person who later continued to chew on the tooth surfaces.[90] Using the flint drills that are similar to the ones found in Mehrgarh and using a bow-drill, the team of Coppa *et al.* could drill a hole in human enamel in less than a minute. When Andrea Cucina, a researcher from the University of Missouri-Columbia, looked the cavities using an electron microscope, he found that the sides of the cavities were perfectly round to be caused by bacteria. He also noticed concentric groves left by the drill. According to him, some plants or other substances were put into the hole to prevent any further decay in the molars. These fillings were decayed during the 9,000 years of seasoning.[91]

8.3.1 PLASTIC SURGERY

In *Rāmāyaṇa*, attractive and voluptuous Sūrpaṇakhā, sister of Rāvaṇa, tried to devour Lord Rāma, a married man. Lord Lakṣmaṇa, younger brother of Lord Rāma, decided to cut off her nose and earlobes as punishment. King Rāvaṇa took care of the problem by asking his surgeon to reconstruct the nose and earlobes of his sister.[92] In time, nasal amputation also worked its way into the metaphors and the Hindi term *Nāk kat gai* (nose is chopped) implies that a person is insulted. Also, "saving nose" (*nāk bacā lī*) is a colloquial term meaning to go through difficult circumstances without embarrassment.

Suśruta described the technique to graft skin, presently known as "plastic surgery" as a general umbrella term. He repaired the nose or earlobes using an adjacent skin flap. Today, this procedure is popularly called as "the Indian method of rhinoplasty"[93] although no plastic is used in the process. Live flesh from the thigh, cheek, abdomen, or the forehead was used to make the new artificial parts.

This procedure was not practiced in the West until the second half of the 15th century in Sicily, an empire with considerable contact with Arabia.[94] In England, the first article on rhinoplasty appeared in the *Gentleman's Magazine* in 1794, written by Colly Lyon Lucas, a British surgeon and member of the Medical Board at Madras, India.[95] He described the process in

[88]Coppa *et al.*, 2006
[89]Coppa *et al.*, 2006
[90]Coppa *et al.*, 2006
[91]Coppa *et al*, 2006.
[92]Brain, 1993.
[93]Sorta-Bilajac and Muzur, 2007.
[94]Pothula, 2001.
[95]Brain, 1993; Lucas, *The Gentleman's Magazine*, 64, pt. 2, no. 4, 891–92, October, 1794. In this article, only the initials of the author are provided.

his letter to the Editor, describing the process as "long known in India" and not known to the British. Colly Lyon Lucas witnessed a case where a local Indian serving the British army, in the war of 1792 CE, was captured by King Tipu Sultan. Unable to defeat the British outright, the sultan tried to starve his enemies by ambushing the Indian bullock drivers who transported grain to the British. Tipu decided to humiliate the bullock drivers by mutilating their noses and ears. Lucas describes one such victim of this practice, the Mahratta bullock driver Cowasjee, who, on his capture, had his nose and one of his hands amputated by the sultan. After one year, this man decided to get his nose repaired. An operation was performed by the Indian doctors who took skin from the forehead and placed it on the nose in proper form. The whole process took about 25 days. The artificial nose looked "as well as the natural one" and the scar on the forehead was not very observable after a "length of time." The procedure generated great interest among the British surgeons. Lucas writes:[96] "For about 12 months he remained without a nose, when he had a new one put on by a man of the Brickmaker cast near Poonah [Poona]. This operation is not uncommon in India, and has been practiced from time immemorial.. . . A thin plate of wax is fitted to the stump of the nose, so as to make a nose of good appearance. It is then flattened, and laid on the forehead. A line is drawn round the wax, and the operator then dissects off as much skin as it covered, leaving undivided a small slip between the eyes. This slip preserves the circulation till an union has taken place between the new and old parts. The cicatrix of the stump of the nose is next pared off, and immediately behind this raw part an incision is made through the skin, which passes around both *alae*, and goes along the upper lip. The skin is now brought down from the forehead, and, being twisted half round, its edge is inserted into this incision, so that a nose is formed with a double hold above, and with its alae and septum below fixed in the incision. A little Terra Japonica is softened with water, and being spread on slips of cloth, five or six of these are placed over each other, to secure the joining. No other dressing but this cement is used for four days. It is then removed and cloths dipped in ghee (a kind of butter) are applied.. . . For five or six days after the operation, the patient is made to lie on his back; and on the tenth day, bits of soft cloth are put into the nostrils, to keep them sufficiently open. The artificial nose is secure, and looks nearly as well as the natural one; nor is the scar on the forehead very observable after a length of time."

In England, The first operation of rhinoplasty was performed by Joseph Constantine Carpue on October 23, 1814, in front of a large group of surgical colleagues and his students. Carpue performed the second operation on an army officer who lost his nose during the Peninsular War against Napoleon, and later wrote a monograph about it.[97]

In his book, *A Short History of Medicine*, the author Erwin Heinz Ackernecht states that "[t]here is little doubt that plastic surgery in Europe, which flourished first in medieval Italy, is a direct descendant of classic Indian surgery."[98] A well known European surgeon, who restored

[96]taken from Ang, 2005.

[97]Brain, 1993; Carpue, 1816; a brief summary of the history of this procedure in India is provided by Ang, 2005; Rana and Arora, 2002.

[98]Ackernecht, 1982. p. 41.

a lost nose, was Branca de Branca from Sicily, using Suśruta's adjacent flap method. His son, Antonio Branca, used tissue from the upper arm as the reparative flap in his operations (around 1460), and "the Italian method" using a distant flap was born. The method was most extensively described by Gaspare Tagliacozzi in his *Chirurgia curtorum* in 1597,[99] almost two centuries prior to when the British learned.

The following procedure was provided by Suśruta in the repair of a nose: "The portion of the nose to be covered should be measured with a leaf. A piece of skin of the required size should then be dissected from the cheek, and turned back to cover the nose. The part of the nose to which this skin is to be attached or joined, should be made raw, and the physician should join the two parts quickly but evenly and calmly, and keep the skin properly elevated by inserting two tubes in the position of nostrils, so that the new nose may look normal. When the skin has been properly adjusted a powder composed of licorice, red sandal-wood, and extract of barberry should be sprinkled on the part. It should be covered with cotton, and white sesame oil should be constantly applied. The patient should take some clarified butter. When the skin has united and granulated, if the nose is too short or too long, the middle of the flap should be divided and an endeavor made to enlarge or shorten it."[100]

The *Suśruta-Saṁhitā* demonstrated a procedure to mend an earlobe with a patch of skin-flap scraped from the neck or the adjoining parts. "The modes of bringing about an adhesion of the two severed parts of an ear-lobe are innumerable; and a skilled and experienced surgeon should determine the shape and nature of each according to the exigencies of a particular case."[101]

"A surgeon well-versed in the knowledge of surgery should slice off a patch of living flesh from the cheek of a person devoid of ear-lobes in a manner so as to have one of its ends attached to its former seat. Thus, the part, where the artificial ear-lobe is to be made, should slightly scarified (with a knife), and the living flesh, full of blood and sliced off as previously directed, should be attached to it (so as to resemble a natural ear-lobe in shape)."[102]

Two thousand years after Suśruta, this operation essentially follows the same procedure. In these operations, "the Indians became masters in a branch of surgery that Europe ignored for another two thousands years," acknowledges Majno, a well-known historian of medicine.[103] Ilza Veith and Leo M. Zimmerman, in their book *Great Ideas in the History of Surgery*, make a similar conclusion: "It is an established fact that Indian plastic surgery provided basic pattern for Western efforts in this direction."[104]

[99] Sorta-Bilajac and Muzur, 2007.
[100] *Suśruta-Saṁhitā*, *Sūtrasthānam*, 16: 46–51.
[101] *Suśruta-Saṁhitā*, *Sūtrasthānam*, 16: 25.
[102] *Suśruta-Saṁhitā*, *Sūtrasthānam*, 16: 4.
[103] Majno, 1975, p. 291.
[104] Veith and Zimmerman, 1993, p. 63.

8.3.2 CATARACT SURGERY

The earliest form of cataract surgery, now known as 'couching,' was first introduced by the ancient Hindus. Suśruta explains the procedure in which a curved needle is used to push the opaque phlegmatic matter in the eye out of the way of vision.[105] Immediately after the surgery, the eye is sprinkled with breast milk and clarified butter.

This procedure of the removal of cataract by surgery was also introduced into China from India. Two poets from the Tang dynasty, Bo Juyi (772–846 CE) and Liu Yuxi (772–842 CE), wrote about a brahmin removing the cataract using a golden probe.[106] According to Guido Majno, a professor of pathology from the University of Massachusetts, the cataract operation described by Aulus Cornelius Celsus (25 BCE - 50 CE) is perhaps derived from Suśruta.[107] Celsus was a celebrated Roman medical writer and physician. His book *De Medicina* was a standard book of medicine in Rome.

8.3.3 CARPENTER ANTS SUTURING AND LEECH THERAPY

Suśruta used somewhat unconventional from the modern standards but highly successful methods to take care of some surgical issues. In the treatment of intestinal surgery to remove obstructing matter, Suśruta suggested to "bring the two ends of intestines that needed to join together. The intestines should be firmly pressed together and large black ants should be applied to grip them quickly with their claws. Then the bodies of the ants having their heads firmly adhering to the spots, as directed, should be severed and the intestines should be gently reintroduced into the original position and sutured up."[108] This may be a crude method for some. However, it was effective procedure and worked well.

The leeches were used to suck pus and blood in the cure of boils, tumors, and other similar diseases by Suśruta. He defines the types of leeches that should be used and provided details of the process. Leeches worked well when the patients were not fit for operations and bleeding was an issue since, with leeches, almost no quality blood was lost.[109] Suśruta first describes twelve different kinds of leeches and suggest to use only six types for the purpose of human treatment. Suśruta suggests that the "leeches should be caught with a piece of wet leather, or by some similar article, and then put in to a large-sized new pitcher filled with the water and ooze or slime of a pool. Pulverized zoophytes and powder of dried meat and aquatic bulbs should be thrown into the pitcher for their food, and blades of grass and leaves of water-plants should be put into it for them to lie upon. The water and the edibles should be changed every second or third day, and the pitchers should be changed every week."[110] "The affected part [of the patient] should be sprinkled over with drops of milk or blood, or slight incisions should be made into it

[105] *Suśruta-Saṁhitā*, Uttaratantra, 17–69: 55 , verses 55–69.
[106] Deshpande, 2008.
[107] Majno, 1975, p. 378.
[108] *Suśruta-Saṁhitā, Cikitsāsthānam*, 14: 20–21.
[109] *Suśruta-Saṁhitā, Sūtrasthānam*, Chapter 13
[110] *Suśruta-Saṁhitā*, Sūtra-sthāna, 13: 15.

in the event of their refusing to stick to the desired spot. . . while sucking, the leeches should be covered with a piece of wet linen and should be constantly sprinkled over with cold water. A sensation of itching and of a drawing pain at the seat of the application would give rise to the presumption that fresh blood was being sucked, and the leeches should be forthwith removed. Leeches refusing to fall off even after the production of the desired effect, or sticking to the affected part out of their fondness for the smell of blood, should be sprinkled with the dust of powered rock salt."[111]

To use these leeches again, Suśruta suggests that the "leeches should be dusted over with rice powder and their mouths should be lubricated with a composition of oil and common salt. Then they should be caught by the tail-end with the thumb and the forefinger of the left hand and their backs should be gently rubbed with the same fingers of right hand from tail upward to the mouth with a view to make them vomit or eject the full quantity of blood they had sucked from the seat of the disease. The process should be continued until they manifest the fullest symptoms of disgorging."[112]

Using leeches for surgical procedures is not just an ancient practice. In 2004, the Food and Drug Administration in America cleared the use of leeches (*Hirudo medicinalis*) as a "medical device" appropriate for certain procedures. Today, surgeons use leeches often in procedures that require skin grafting or regrafting amputated appendages, such as finger or toes. If the veins do not flow blood to the damaged region, leeches are used to ooze the blood so that the body's own blood supply eventually gets established and the limb survives.[113] Also, the saliva of leeches contains an anti-clotting agent, called hirudin, which allows the blood to flow freely and avoid congested skin flaps problem. It also has hyaluronidase, histamine-like vasodilators, collagenase, inhibitors of kallikrein and superoxide production, and poorly characterized anaesthetic and analgesic compounds. Thus, the saliva is analgesic, anaesthetic and has histamine-like compounds.[114] The therapy is pretty inexpensive, with each session taking about 40 minutes and leeches worth $7–10.[115] For each session, a new group of leeches are used and dumped as infectious waste material after the treatment. In Germany alone, about 350,000 leeches were sold in 2001.[116] Their use is even more prevalent now.

In a study conducted at the Department for Internal and Integrative Medicine, University of Essen, Germany, sixteen people in the test group were recruited who endured persistent knee pain for more than six months and had definite radiographic signs of knee osteoarthritis. Ten out of sixteen proceeded with leech therapy and avoided conventional treatment while the remaining six tried only the conventional treatment. In all patients with leech therapy, no adverse effect or local infection was observed. Some patients described the initial bite of leeches as painful. It was

[111] *Suśruta-Saṁhitā, Sūtrasthānam*, 13: 18.
[112] *Suśruta-Saṁhitā, Sūtrasthānam*, 13: 19–20.
[113] Michalsen *et al.*, 2001; Michalsen *et al.*, 2006; Rados, 2004.
[114] Michalsen et al., 2001; Oevi *et al.*, 1992.
[115] Rados, 2004.
[116] Michalsen *et al.*, 2001.

found that the group with leech therapy did considerably better than the group with conventional therapy. Their pain was reduced significantly after three days and the good effects lasted up to four weeks. Later, a second group with 52 people was was experimented with similar therapy that yielded similar results.[117]

8.4 HINDU MEDICINE IN OTHER WORLD CULTURES

The Hindu medicine was popular in China, the Middle East, and Europe from the ancient period. Ktesias, a Greek physician who lived in the Persian court during the early fifth-century BCE for some 17 years, wrote that the Indians did not suffer from headaches, or toothaches, or ophthalmia.[118] They also did not have "mouth sores or ulcers in any part of their body."[119] Claudius Aelianus (175–235 CE), who lived during the period of Emperor Septimus Severus in Rome, preferred drugs from India in comparison to the Egyptian drugs. He wrote: "So let us compare Indian and Egyptian drug and see which of the two was to be preferred. On the one hand the Egyptian drug repelled and suppressed sorrow for a day, whereas the Indian drug caused a man to forget his trouble for ever."[120] This statement implies that the Indian herbs and drugs were well received by the Romans.

Al-Bīrūnī (973–1050 CE) mentions the availability of an Arabic translation of *Caraka-Saṁhitā* in the Middle East.[121] As Arab medicine became popular in Europe, the name of Caraka is repeatedly mentioned in medieval-Latin, Arabic, and Persian literature on medicine.[122] Al-Jāḥiz (ca. 776–868 CE), a Muslim natural philosopher from Arabia, acknowledged that the Hindus possess a good knowledge of medicine and "practice some remarkable forms of treatment."[123] "They were the originators of the science of *fikr*, by which a poison can be counteracted after it has been used."[124]

Ṣā'id al-Andalusī (d. 1070 CE) of Spain also acknowledged the vast knowledge of the Hindus in medicine. "They [Hindus] have surpassed all the other peoples in their knowledge of medical science and the strength of various drugs, the characteristics of compounds and the peculiarities of substances."[125] Nearby in England, Roger Bacon (1214–1292), a twelfth century natural philosopher, wrote a book for the Pope and noticed that the Indians "are healthy without infirmity and live to a great age."[126] Marco Polo, an Italian traveler, also noticed the long lives of the Hindus. "[T]hese Brahmins live more than any other people in the world, and this comes about through little eating and drinking, and great abstinence which they practise [practice]

[117] Moore and Harrar, 2003.
[118] McCrindle, 1973, p. 18.
[119] McCrindle, 1973, p. 18.
[120] Aelianus, 1958, 4: 41.
[121] Sachau, 1964, p. XXVIII.
[122] Royle, 1837, p. 153.
[123] Pellat, 1969, p. 197.
[124] Pellat, 1969, p. 197.
[125] Salem and Kumar, 1991, p. 12.
[126] Bacon, 1928, *Opus Majus*, p. 372.

more than any other people. And they have among them regulars and orders of monks, according to their faith, who serve the churches where their idols are, and these are called yogis, and they certainly live longer than any others in the world, perhaps from 150 years to 200."[127] It is likely that Marco Polo was overstating their longevity in his reference to brahmins living upto to 150 to 200 years. However, the point remains. The Hindus were known to have lived long lives.

"Hindoo [Hindu] works on medicine having been proved to have existed prior to Arabs, little doubt can be entertained," writes John Forbes Royle (1798–1858), in his book *Antiquity of Hindoo Medicine*, to advocate the antiquity of Hindu medicine and its role in modern medicine.[128] "We can hardly deny to them [Hindu] the early cultivation of medicine; and this so early, as, from internal evidence, to be second, apparently to none with whom we are acquainted. This is further confirmed by the Arabs and Persians early translating of their works; so also the Tamuls [Tamils] and Cingalese [Singhalese, people of Sri-Lanka] in the south; the Tibetans and Chinese in the East; and likewise from our finding, even in the earliest of the Greek writers, Indian drugs mentioned by corrupted Sanscrit [Sanskrit] names. We trace them at still earlier periods in Egypt, and find them alluded to even in the oldest chapters of the Bible."[129] Royle served as surgeon for the East India Company and lived in India. His interest in traditional botanical Hindu remedies of various diseases led him to write his classic book.

Abu Yusuf Ya'aub ibn Ishaq al-Kindī, popularly known as al-Kindī (ca. 800–870 CE), is considered to be the "philosopher of Arabia." He was born and lived in Baghdad. He wrote a medical formulary, called *Aqrābādhīn*, which was translated by Martin Levey into English.[130] While studying the *materia medica* defined by al-Kindī, Levey concludes that about 13 percent comes from the Indian region. In his view, "many of the Persian *materia medica* may more properly be considered to be Indian."[131] In that case, the Indian-Persian *materia medica* in al-Kindī's work turns out to be about 31 percent.[132]

Alī B. Rabban Al-Ṭabarī (783–858 CE), a Persian physician from Tehran, who later lived in Baghdad serving Caliph al-Mutawakkil (reigned 847–861 CE) as his physician, wrote an encyclopedic book on medicine, *Firdaws al-Ḥikmah* (*Paradise of Wisdom*).[133] It was translated into English by Muḥammad Zubair Siddiqi in 1928.[134] This book contains some thirty-six chapters on Hindu medicine and refers to the works of noted Indian physicians/thinkers such as Caraka, Suśruta, Cāṇakya (Kauṭilaya), Mādhavakara, and Vagbhata II. Al-Ṭabarī mentions three *doṣa* and seven *dhātu* of ayurveda for medical treatments. This is in accordance with the tradition of ayurveda in India.

[127] Needham, 1981, p. 81.
[128] Royle, 1837, p. 62.
[129] Royle, 1837, p. 152.
[130] al-Kindī, 1966.
[131] al-Kindī, 1966, p. 20.
[132] Levey and al-Khaledy, 1967, p. 33.
[133] see Meyerhof, 1931; *Dictionary of Scientific Biography*, 13: 230.
[134] Siddiqi, 1928.

Max Meyerhof has written excellent accounts of al-Ṭabarī's work and connected it with *Suśruta-Saṁhitā*, *Caraka-Saṁhitā*, the *Nidāna* and the *Aṣṭāṅgahṛdava-Saṁhitā*. The *Nidāna* is a work on pathology written by Mādhavakara. This work was translated into Arabic under the patronage of Hārun al-Rashīd. The *Aṣṭāṅgahṛdava-Saṁhitā* is the work of Vagbhata II.[135] Max Meyerhof concludes the existence of Indian drugs in Arabic treatises: "We find indeed, in the earliest Arabic treatises on medicine, the mention of many Indian drugs and plants which were unknown to Greeks, and all of them bearing Sanskrit names which had passed through the New Persian language."[136]

Triphalā is a popular drug in India. The word *Triphalā* means "three-fruits" since the drug is made from three ingredients: *haritaki* or simply *har* (*Terminalia chebula*), *bahera* (*Terminalis belerica*), and *āmalaki* (*Phyllanthus emblica*). The Arabs used the term *atrifal* for *triphalā* that is similar in pronunciation, while the Chinese literally translated the term and called it *san-teng*, implying three herbs.[137] These literary evidences are examples of the transmission of medical knowledge from India to Arabia and China.

In studying the transmission of ayurvedic knowledge to China, Chen Ming, a Chinese scholar, suggests that "scholars of medical history have long been well aware of the Indian influence on Chinese medicine."[138] In providing examples of such knowledge, he cites the division of medicine into eight branches in Chinese books, as done by Caraka and Suśruta. Dharmakṣema, a scholar, in his book *Daban nie jing*, followed the same division in 421 CE. Paramārtha (499–569 CE) translated a Sāṁkhya text, *jinqishilum*, that refers to eight divisions of medical remedies. Sun Simiao (581–682), a medical writer and physician for the Sun and Tang dynasties, wrote about ayurvedic practices in his work *Beiji Qianjin yao fang* (Important medicinal formulae [worth] a thousand [pieces of] gold). He mentions Jīvaka's story, discussed in Section 8.1, that all plant life has medicinal value. Simiao is known for the text "On the Absolute Sincerity of Great Physicians," popularly known as the Chinese Hippocratic Oath, written in the first chapter of the above-mentioned book. This oath is still required reading for Chinese physicians. Simiao also attributed several medicinal formulations to Jīvaka: Jīvaka's ball medicine for ten thousand illnesses, Jīvaka's medicines for illnesses caused by evil spirits, Jīvaka's soup, Jīvaka's medicine for prolonging life without getting old. He introduced "*chan*" or "*dhyāna* (meditation) and yoga into China, and also introduced a particular massage technique and called it Brāhmin's method.[139]

Tombs in Turfan (also Turpan, 42°59′N, 89°11′E, near the Silk Road) yield medical fragments in Sanskrit and Tocharian, the local language. Fragments of the ayurvedic texts, *Bhela Saṁhitā*, *Siddhasāra*, and *yoga-śataka*, have been found in these tombs. The epitaph of Lüsbuai (Battalion Commander) Zhang Xianghuan (681 CE) mentions two ayurvedic physicians, Jī-

[135] Meyerhof, 1931 and 1937.
[136] Meyerhof, 1937.
[137] Mahdihassan, 1978.
[138] Ming, 2006.
[139] Deshpande, 2008.

vaka and Nāgārjuna. In his studies, Chen Ming concludes that "Indian medicine also influenced medical practices in Medieval Turfan."[140]

Su Jing in the seventh century revised the Tang pharmacopoeia called or *Xinxiu bencao* (Newly revised materia medica) and noted down a remedy for Beriberi. He defined the prescription as 'Brahmin's prescription'. In 752 AD, Wang Tao of the Tang dynasty wrote a large compendium, *Waitai miyao fang* (Medical secrets of an official). This book again records a treatment of beriberi using the 'Brahmin's prescription'. Both of these records suggest that a successful cure for Beriberi came to China from India.[141]

[140] Ming, 2006.
[141] Deshpande, 2008.

CHAPTER 9

The Global Impact

The preceding chapters covered the multi-faceted concepts of Hindu science: numeral system with its place-value notations; mathematical processes that cover arithmetic, algebra, and trigonometry; the shape of the Earth and planetary motions; the constitution of matter; the properties of matter; standards for mass, length and time; and physical and chemical processes involved in the making of drugs, poisons, and new compounds, so-called plastic surgery, rust-free Iron Pillar, yoga, etc. These chapters basically cover most branches of modern science and provide the substantial contributions that the ancient Hindus made to science. Also, as mentioned in Chapter 1, the rationale of why the American Association for the Advancement of Science (AAAS) decided to credit the Hindus for their knowledge in mathematics and astronomy is also provided. These chapters also provided the global impact of specific discoveries and inventions of the ancient Hindus. The current chapter focuses only on some additional new information.

If we need to mention just one thought that is central to Hinduism and which could play an important role in the future of humanity as it did in the past, it is a hymn of *Ṛgveda*: "To what is One, sages call by different names."[1] The underlying meaning is that God is one and people call Him by different names. Though this doctrine is not a part of science, it is perhaps the most timely message in view of the religious strife around the world. Al-Bīrūnī, an Islamic philosopher, understood this well when he wrote: "The Hindus believe with regard to God that he is one, eternal, without beginning and end, acting by free will, almighty, all-wise, living, giving life, ruling, preserving, . . ."[2] It is due to the impact of this doctrine that the Hindus did not subjugate other religions and tried to spread their own message by winning the hearts of people. Christians, Jews, Zoroastrians, and Muslims found refuge in India at different times in history and could flourish there. Out of the currently 2.6 million population of Zoroastrians in the world, most live in India, a country of refuge. In contrast, elsewhere in world history, there have been more wars in the name of religion than on anything else.

For the ancient (and modern) Hindus, exploring the truth, including science, was considered helpful in achieving liberation (*mokṣa*). Therefore, scientific investigations were encouraged through religious codes, as explained in Chapter 2. Long before the Christian era, the ancient Hindus had established an educational system which was comparable to the present university system. The intellectual centers of Nālandā (near Patna, India), Taxila (Takṣaśilā, near Islam-

[1] *Ṛgveda*, 6: 22: 46.
[2] Sachau, 1964, vol. 1, p. 27.

abad, Pakistan), Kānchipuram (Tamilnadu, India), Vikramśilā (Bihar, India), Varanasi (Uttar Pradesh, India), and Valabhi (Gujrat, India) were perhaps the most well known during the ancient period. Taxila was visited by Alexander the Great and later supported by Aśoka. It had a large number of *stūpas* (pillars) and monasteries. The center of Taxila was destroyed during the fifth century by the Huns. Nālandā (near Patna, Bihar) is the place where Nāgārjuna and Āryabhaṭa I taught. Nālandā was visited by three famous Chinese visitors: Faxian (or Fa-Hien), Xuan Zang (or Hiuen Tsang), and Yijing (or I-tsing). At one time, it had about 10,000 students, 1,500 teachers, and 300 classrooms. It was completely destroyed in 1193 by the army of Bakhtiyar Khilji. The library was set on fire and, due to the extensive collection of books, it took over three months for the fire to be extinguished. Other intellectual centers, such as in Varanasi, destroyed in 1194 CE by Qutb-ud-din Aibak, and Mathura, destroyed in 1018 CE by Mahmud of Ghazni, also suffered.[3]

Substantially greater effort is required in order to unearth and fully comprehend the foundational stones of a vast intellectual treasure that we owe to the ancient Hindus. For some reason, the scholarship of the last century on Hindu science is scant and has not received the interest of scholars as it did during the nineteenth century. There are only a small number of courses on Hinduism in the West, while courses that focus on the anthropological studies of India and sensational issues, such as *sati*, the caste-system, eroticism in literature, human trafficking, etc. have been steadily growing. At the time of this writing, there is not even a single course in America that deals primarily with Hindu science. In contrast, the ancient Greeks, whose contributions to science were comparable to those of the Hindus, are covered in various courses in most universities. The main reason is that most academic institutions are shaped by the interests of donors. It may sound ridiculous to some; however, it is the brutal reality of academia. Princeton University offered 13 courses just on Greece during the Spring 2008 semester. This is in part due to Stanley J. Seeger's endowment to the university.[4] Similar endowments for Hinduism are lacking and there is little or no support from the governments of various countries. This is not the case with other religions.

Most of the knowledge of the ancient Hindus was transferred to the West via the Islamic domination of Europe. This transfer of knowledge is reminded by the Sanskrit words that are now a part of the English language in original or metamorphosed form. A list of these words makes a strong case for the Hindu science, as they are related to the scholarly tradition of the ancient Hindus. Numbers (two, three, six, seven, eight, and nine), pundit, guru, read, sine (as a trigonometric function), geometry, zero, sulfur, gold, silver, lead, uterus, yoga are some of the

[3]I have decided to avoid any description of the destruction that the ancient Hindus and their institutions had to deal with. The destruction of Baghdad by Hulagu Khan, the destruction of Spain during the Inquisition, and the destruction of Alexandria by a religious zealot group are some examples of how a country or a culture can be subdued. Most examples of destruction cited here outside the Indus-Valley region occurred for a short period. In contrast, the Hindus faced subjugation and destruction of their intellectual centers for nearly a millennium. It is magical the way India, with majority Hindu population, has bounced back in science and technology even after many centuries of foreign rule during Mogul and colonial period that tried to destroy the educational system of Hindus and extinguish their spirit to search for truth.

[4]Arenson, 2008.

examples of the Sanskrit words in English.[5] All cultures have words to define a scholar in their languages. However, it is the Sanskrit word "guru" or "pundit" that has prevailed in English.

9.1 IMPACTS DURING THE ANCIENT AND MEDIEVAL PERIODS

During the ancient and medieval periods, the Hindu corpus attracted the great minds from all over the world, such as: Pythagoras, Apollonius of Tyana, and Plotinus from the Greek or Roman tradition; al-Bīrūnī, al-Fazārī, Ibn Sīnā, al-Jāḥiz, al-Khwārizmī, al-Mas'udī, Kūshyār Ibn Labbān, and al-Uqlidīsī from the Islamic tradition; Faxian (Fa-Hien), Xuan Zang (or Hi-uen Tsang), and Yijing (or I-tsing) from the Chinese tradition; and Pope Sylvester, Ṣā'id al-Andalusī, Roger Bacon, Adelard of Bath, and Leonardo Fibonacci from the European tradition.

The ancient Hindus invented the game of chess, called *caturaṅga* in Sanskrit. The word signifies "four members of an army"—elephants, horses, chariots, and foot soldiers. It was called *al-śatranj* in Arabic. In the game, a key move is made to trap the king, called *eschec* in French and "check" in English. This led to the word chess for the game. The game epitomizes the science of warfare that helped the ancient and medieval strategists in warfare. The role of horses, elephants, foot-soldiers, chariots, and the king are well defined in a set configuration. The main task was to protect your own king and kill the king of your enemy. Ṣā'id al-Andalusī, (died in 1070 CE) of Spain considered the discovery of chess as a testimony to the "clear thinking and the marvels of their [Hindu] inventions. . . . While the game is being played and its pieces are being maneuvered, the beauty of structure and the greatness of harmony appear. It demonstrates the manifestation of high intentions and noble deeds, as it provides various forms of warnings from enemies and points out ruses as well as ways to avoid dangers. And in this there is considerable gain and useful profit."[6] Today, chess is played all over the world and international tournaments are conducted to select a world champion.

9.1.1 IMPACT ON ARABIA

The work of al-Khwārizmī in Baghdad was made possible due to the discoveries and inventions of the ancient Hindus. Al-Khwārizmī work led to the inclusion of several new words in the English dictionary: algebra, zero, algorithm, sine function, to name a few. His books were read by established scholars in Europe, such as Copernicus, Adelard of Bath, and Leonardo Fibonacci, either in Arabic or in translation. Al-Khwārizmī's main contributions were to compile Hindu

[5]amrita, Aryan, ashram, atman, acharya, ahimsa, bandana, basmati, candy, cheetah, cot, jackal, jungle, karma, kin, kundalini, lemon, mahout, man, Mandarin, mantra, maya, mix, namaste, nirvana, orange, pajamas, puja, samadhi, sapphire, shampoo, sugar, swastika, tantra, yoga, yogi, and zen are non-science representative words that are of Sanskrit in origin. The followers of Lord Kṛṣṇa celebrate festivals by carrying images of Him in a huge wagon all over India. Lord Kṛṣṇa's another name is Jagan-nāth, the lord of the world. For Europeans, it was a waste of time and money for senseless devotion. Thus, the word juggernaut evolved to define large overpowering and destructive forces or objects. Similarly, thug and dacoit are also Sanskrit in origin. In some case the path of adoption is a bit convoluted.

[6]Salem and Kumar, 1991, p. 14.

knowledge of mathematics and astronomy in his books, as indicated from the title of his books. It was the Hindu mathematical tools that allowed Copernicus to figure out his heliocentric solar system.

Copernicus knew some mathematical tools of the ancient Hindus that he may have learned during his stay in Italy where people like Pope Sylvester and Leonardo Fibonacci had already documented the mathematics of the Hindus. As mentioned earlier, Copernicus opted to use Hindu arithmetic in writing his book, *De revolutionibus orbium coelestium* (On the Revolutions of the Heavenly Spheres). Copernicus is not the only celebrated scientist who was benefited with the contributions of the ancient Hindus; the table of parallax of the Moon that was suggested by Johannes Kepler (1571–1630) was identical to the one given by Brahmgupta in *Khaṇḍa-Khādyaka* almost a millennium ago, as inferred by Otto E. Neugebauer in his analysis.[7]

Khalīl wa Dimna is one of the celebrated books in the Islamic world. This book is based on an Indian book, *the Pañca-tantra*, meaning "five-principles." The original Sanskrit book was written before the Christian era by Viṣṇu Sharma to educate the sons of an Indian king on human conduct and the art of governing (*nīti*) using animal fables. Burzuwaih (or Borzuy), a personal physician of Anoushiravan (fl. 550 CE) of the Sassanid dynasty, made a trip to India to collect medicinal herbs for the king. He came back with the book and translated it into the Persian language. After the Islamic conquest of Persia, the book was translated by Ibn al-Muqaffa (8th century CE) into Arabic as *Khalīl wa Dimna* where Khalīl and Dimna are the two jackal characters in several stories in the book. All stories ended with a question and the next story was the answer of the previous question. These stories have ethical, social, and political wisdom. *Khalīl wa Dimna* became popular in the Middle East and was known in Europe by the eleventh century. Ṣā'id Al-Andalusī praised the contents of *Khalīl wa Dimna* in his book *Ṭabaqāt al-'Umam* and correctly labeled Hind as the country of origin. In his view, the book is useful for the "improvement of morals and the amelioration of upbringing" and "is a book of noble purpose and great practical worth."[8]

9.1.2 IMPACT ON CHINA

Mandarin is one of the two most prominent languages in China and sets the norms of aristocratic communication—a language used by the upper class. It was the spoken language of the bureaucrats who were courteous and polite in their behavior. The word itself is derived from the Sanskrit word *mantrin*, meaning a minister or councilor. Even today in India, all cabinet ministers are called *mantrī* and the Prime-minister is called *Pradhān-mantrī*, meaning the prime councilor. The Sanskrit language influenced the aristocrats of China who adopted the language. As a result, some Sanskrit words entered in the Chinese language. For example, *kṣaṇa* (a moment), *śarīra* (body) and nirvana in Sanskrit became *chānā*, *shélizi*, and *niepān*, respectively, in Chinese. Of course, the most significant example of this influence is the fact that Buddhism

[7]Neugebauer, 1962, p. 124.
[8]Salem and Kumar, 1994, p. 14.

became the most prominent religion in China without any war or conflict. This is in contrast to several other nations where people had to choose between their lives or their religion. Such was the influence of India on China.

9.1.3 IMPACT ON GREEK SCIENCE AND PHILOSOPHY

Cyrus the Great (558–530 BCE) of Achaemenid Dynasty ruled all the way from Greece to the Indus River. Being part of a common empire, the Greeks learned about India. In 327 BCE, Alexander the Great fought back, regained Greece from Persia, and later controlled Persia and India. When the army of Alexander could go through hostile conditions to reach India, it was also possible for a lone scholar to visit India too. The historical documents provide information about the influence of Indian philosophy on two prominent Greek or Roman philosophers.

Pythagoras

Pythagoras advocated vegetarianism, metempsychosis (transmigration of the soul from one physical body to another, just as we cast off old clothes and wear new one), fasting as a way of purification, chastity as a virtue, and the sanctity of animal life that must be honored. He promoted orality in the preservation of knowledge, monism as an ideology, and believed on the existence of a soul in plants, animals, and humans. The above-mentioned doctrines of Pythagoras were essential to his school, and were practiced by his followers in the West for almost a millennium. These doctrines were new to Greece during the period of Pythagoras. Elsewhere, there is only one place in the world where all the above-mentioned doctrines existed prior to Pythagoras. This is the land of the Hindus, the Indus-Sarasvatī region. As is the case with the Hindu tradition, Pythagoras did not write any book since he contended that knowledge should be veiled from undesirable people; only oral communication was used to transfer knowledge.

Apollonius of Tyana (15–100 CE), a noted Pythagorean philosopher, learned the doctrine of nonviolence or the sanctity of human and animal life in the tradition of Pythagoras who, in Apollonius' view, was himself taught by Hindu philosophers. Philostratus quotes the words of Apollonius: "I did not sacrifice, I do not: I do not touch blood, even on the alter, since that was the doctrines of Pythagoras. . . it was the doctrines of the Naked Philosophers of Egypt and the Wise Men of India, from whom Pythagoras and his sect derived the seeds of their philosophy."[9]

Apollonius correctly considered the Indian civilization to be much more ancient than the period of Pythagoras. "Pythagoras was anticipated by the Indians, lasts not for brief time, but for an endless and incalculable period," opines Apollonius.[10] Apollonius was a person of repute, visited India, and was compared with Jesus of Nazareth by some Christians during the early centuries of Christianity.

Titus Flavius Clemens (c.150–215 CE), popularly known as Clement of Alexandria, was a Christian theologian who was familiar with classical Greek philosophy and literature. He

[9]Book 8, Chapter 7, 12; Philostratus, 1960, vol. 2, p. 339.
[10]Philostratus, 1960, vol. 2, p. 49.

believed that the Greek philosophy had non-Greek origins. Similar to Apollonius, Clement, in his book *Stromata* (*The Miscellanies*), writes that Pythagoras went to Persia and came in contact with the Brahmins. In his view, Pythagoras learned philosophy from his interactions with these brahmins: "Pythagoras was a hearer of . . . the Brahmins."[11]

Eusebius of Caesarea (263–339 CE), a Roman Christian historian, in his book *Praeparatio Evangelica* (*Preparation for the Gospel*) mentions Pythagoras' visit to Babylon, Egypt and Persia. Similar to Apollonius of Tyana and Clement of Alexandria, Eusebius also shares the opinion that Pythagoras "studied under the Brahmans [or Brahmins]," and learned geometry, arithmetic, and music from these foreign lands. Eusebius is also quite clear in his statement that Pythagoras learned "nothing" from the Greek philosophers and "became the author of instruction to the Greeks in the learning which he had procured from abroad."[12]

The visit of Pythagoras to India was conclusive in the mind of Voltaire (1694–1778) (pen name of Francois Marie Arouet) as he writes that "Pythagoras, the gymnosophist, may alone serve an incontestable proof that true science was cultivated in India. . . . It is even more probable that Pythagoras learned the properties of the right-angled triangle from the Indians, the invention of which was afterward ascribed to him."[13] According to Voltaire, "The Orientals, and particularly the Indians, treated all subjects under the veil of fable and allegory: for that reason Pythagoras, who studied among them, expresses himself always in parables."[14] Voltaire had no doubt about Pythagoras' visit to India and wrote: "All the world knows that Pythagoras, while he resided in India, attended the school of Gymnosophists and learned the language of beasts and plants."[15]

D. E. Smith, a noted historian who is known for his classic book, *History of Mathematics*, points out a resemblance between the Hindu and Pythagorean philosophies: "In spite of the assertions of various writers to the contrary, the evidence derived for the philosophy of Pythagoras points to his contact with the Orient. The mystery of the East appears in all his teaching . . . indeed his [Pythagoras'] whole philosophy savors much more of the Indian than of the Greek civilization in which he was born."[16]

On a possible Indian influence on Pythagoras in comparison to Egyptian influence, H. W. Rawlinson (1810–1895 CE) concludes: "It is more likely that Pythagoras was influenced by India than by Egypt. Almost all the theories, religious, philosophical, and mathematical, taught by the Pythagoreans were known in India in the sixth century B.C. [BCE]."[17] Rawlinson is the person who first decoded cuneiform language of Babylon after discovering the Darius' Behistun inscriptions. He also served the British empire and lived in India.

[11] *Stromata*, Book 1, Chapter 15.
[12] *Praeparatio Evangelica*, Book 10, Chapter 4.
[13] Voltaire, 1901, vol. 29, p. 174.
[14] Voltaire, 1901, vol. 24, p. 39.
[15] Voltaire, vol. 4, p. 47.
[16] Smith, 1925, vol. I, p. 72.
[17] Rawlinson, p. 5 in the book by Garratt, 1938.

Plotinus

Plotinus (205–270 CE), the founder of the Neo-Platonic school, attempted to visit India to learn from Brahmins while he was in Alexandria. He joined the army of Emperor Gordian III during his expedition to Persia in 242 CE against the Sassanians king, Sapor, in hopes of visiting India. This is obviously a major sacrifice for a philosopher to pick up weapons and fight, just for learning from India. However, he was not the first. Before Plotinus, Pyrrhon pursued a similar dream to visit India by joining Alexander's army. Pyrrhon succeeded while Plotinus failed since Gordian III was assassinated in Mesopotamia, and Plotinus could not fulfill his dream of an Indian expedition.[18] Plotinus moved to Antioch and later to Rome where he spent the rest of his life. Plotinus also did not write any book, following in the footsteps of Pythagoras and Socrates and the Hindus. Similar to the oral tradition of the Hindu, he only gave lectures while living in Rome. His lectures were later compiled by his followers under the title *Enneads*.

Ammonius Sakkas, Plotinus' teacher, was perhaps an Indian since his name is not a typical Greek name, and is similar to a Sanskrit name, where *Śākya* means a monk.[19] Porphyry, a disciple of Plotinus, tells us that Plotinus "acquired such a mastery of philosophy, that he became eager to gain knowledge of the teaching prevailing among the Persians, as also among the Indians."[20] It is pretty clear that association with Ammonius kindled a desire in Plotinus to join Emperor Gordian's military expedition to Persia and India. The influence of the Hindu philosophy on Neo-Platonism has been suggested by scholars over the years. Rawlinson writes: "It certainly appears probable that Neo-Platonism was affected by Oriental [Hindu] philosophy . . . Hence we may suppose that the doctrines it inculcates, – abstinence from flesh, subjection of the body by asceticism, and so on—are derived from Oriental [Hindu] sources."[21] Rawlinson concludes: "the debt of Neo-Platonism to Oriental sources is indisputable . . ."

Plotinus "produced a new philosophical synthesis: Greek rationalism in the service of Oriental mysticism," as suggested by Jean W. Sedlar.[22] In Plotinus' philosophy, the human soul is a part of the world-soul, implying that we all have divinity within us. When Plotinus was close to death, as we know from his student Porphyry, he said, "I am striving to give back the divine which is in me to the divine in the universe."[23] This implies that his individual soul will merge with the cosmic divine soul. This is the core essence of the Hindu philosophy of *Aham brahmāsmi* (I am the God), which is central to the *Vedas* and *Upaniṣads*[24] and is currently propagated by the New Age movement in America.

Thomas McEvilley, in his book *The Shape of Ancient Thought*, explored the Greek works and compared them with Indian thoughts. On the question of exchanges between these two cultures, McEvilley suggests that "[i]t can be argued that Plato was already Indianized through

[18] Sedlar, 1980, p. 292; McEvilley, 2002, p. 550.
[19] Sedlar, 1980, p. 292; Halbfass, 1988, p. 17.
[20] Gregorios, 2002, p. 17; McEvilley, 2002, p. 550.
[21] Rawlinson, 1926, p. 175.
[22] Sedlar, 1980, p. 292.
[23] MacKenna, 1956, sec. 1, p. 1.
[24] Bṛhadāraṇyaka Upaniṣad, 1: 4: 10.

Orphic and Pythagorean influences, and on that basis alone some, at least, of his works cannot be regarded as 'purely Greek thought.' Plotinus, then may have received the Indian influence from Gymnosophists in Alexandria, or from the works of Plato, or both; it comes to the same thing: he was philosophizing in an Indianized tradition. It is not just a question of whether Plotinus' philosophy was derived from India *by him*. Its major outlines, in the view presented in this book, had been derived from India almost a thousand years earlier and handed down through what might be called the Indianized, or Indian-influenced, strand of Greek philosophy, to which Plotinus emphatically belonged. He could, then, and perhaps would, have come up with his model of things without any additional Indian input in his lifetime, though it seems clear, in any case, that he had some."[25]

9.2 IMPACTS DURING THE MODERN PERIOD

The literature of the ancient Hindus continued to attract modern scientists and philosophers: Ralph Waldo Emerson, Johann Wolfgang von Goethe, Johann Gottfried Herder, Aldous Huxley, Carl Jung, Max Müller, Robert Oppenheimer, Erwin Schrödinger, Arthur Schopenhauer, Nikola Tesla, Henry David Thoreau, and Hideki Yukawa, to name a few. These scholars found the origin or validity of their ideas in Hindu literature. "India has had a significant impact upon the manner in which Europe has articulated, defined, and questioned itself and its fundamental and symptomatic concepts of theory, science and philosophy," write Wilhelm Halbfass.[26] This view is supported by the fact that many European philosophers, including Goethe, Jung, Herder, Humboldt, and Schopenhauer studied Hindu books and introduced Hindu concepts in their own works.[27]

Voltaire (Actual name, Francois-Marie Arouet; 1694–1778 CE), a French historian, philosopher, and dramatist, visited India, Egypt, and Arabia during the eighteenth century, and was well versed with the history of these regions. In the view of Voltaire, "As India supplies the wants of all the world but is herself dependent for nothing, she must for that very reason have been the most early civilized of any country. . ."[28] In Voltaire's view, Indian science was more ancient than Egyptian science. The recent excavations in Margarh are proving him right. Voltaire writes, "If the question was to decide between India and Egypt, I should conclude that the sciences were much more ancient in the former [India],"[29] Voltaire is not alone in assigning priority to the Indians in comparison to Egypt. Benjamin Farrington, a noted historian of science, expressed similar views in his analysis of the Pythagorean Theorem. "The degree of this knowledge, and the possibility of this diffusion from a common center [center], are questions that may one day be answered with a confidence that is now impossible. But when the answer is given, if it ever is, perhaps neither Babylon nor Egypt will appear as the earliest exponent of

[25] McEvilley, 2002, p. 550–551.
[26] Halbfass, 1988, p. 159.
[27] Sedlar, 1982 and 1980.
[28] Voltaire, 1901, vol. 29, p. 180.
[29] Voltaire, 1901, vol. 24, p. 41.

civilization. The Nile and the Euphrates may have to yield place to the Indus [India]."[30] This is in reference to the ancient Hindus' *Śulbasūtra* that contains the so-called Pythagorean Theorem long before it was proposed by Pythagoras. "I am convinced that everything has come down to us from the banks of the Ganges, - astronomy, astrology, metempsychosis, etc. . . It is very important to note that some 2,500 years ago at the least Pythagoras went from Samos to the Ganges to learn geometry. . .But he would certainly not have undertaken such a strange journey had the reputation of the Brahmins' science not been long established in Europe," wrote Voltaire in a letter in 1775.

Johann Gottfried Herder (1744–1803), a German philosopher and poet, considered the Ganges region as "the primordial garden" where human wisdom started and was nourished, and the birth place of all languages, the Sanskrit language being the mother. He also considered Sanskrit poetry as the mother of all other poetry, indicating Vedic literature as the source of most other poetry works elsewhere.[31] Herder influenced another famous German philosopher and poet, Johann Wolfgang von Goethe (1749–1832 CE). Goethe's interest in India came with his reading of Kālīdāsa's famous works – *Śakuntalā* and *Meghduta*—and flourished with his interaction with noted philosophers such as Johann Gottfried von Herder.[32] Goethe even tried to learn Sanskrit to read Hindu literature.[33] He tried to visit India, but without success.[34] Goethe's poor health did not allow him to take the arduous sea journey to India. His popular drama, *Faust*, was inspired by his reading of the German translation of *Śakuntalā* by Forster in 1791.[35] He also wrote two ballads dealing with India: "Der Gott und die Bajadere" and "Der Paria."[36]

Carl Gustav Jung (1875–1961), a Swiss psychiatrist and psychoanalyst who as a young man collaborated with Sigmund Freud on human psychology, was greatly influenced with Hindu scriptures. Jung's contributions to psychology, dream analysis, and New Age movement are immense. Freud and Jung developed differences on the role of libido on human personality and that caused a break-up in their collaboration and personal friendship. Jung basically lost his social circle where he had not many colleagues to discuss his research. He sought refuge in the Hindu literature for philosophical ideas. The ancient works of Patañjali influenced Jung greatly. The books of the Hindus not only influenced his thinking but also provided Jung confirming parallels for his independent insights. "In the absence of like-minded colleagues, Hinduism provided him with evidence that his differences with Freud were founded on experiences shared by other human beings, therefore were not simply the products of a deranged mind," writes Harold

[30] Farrington, 1969, p. 14.

[31] Patton, 1994, p. 207–208.

[32] Herder was called German Brahmin by some for his interest in Hinduism. Herder wrote *Gedanken einiger Brahmanen* (*Thoughts of Some Brahmins*) in 1792 which was based on *Hitopadesha* and the *Bhagavad-Gītā*. To read more about Herder and Hinduism, read Ghosh, 1990.

[33] Dasgupta, 1973, p. 21.

[34] Steiner, 1950, p. 1.

[35] Remy, 1901, p. 20.

[36] Remy, 1901, p. 20.

Coward.[37] Jung wrote two articles that directly dealt with his impression of India: "The Dream-like World of India" and "What India Can Teach Us." Both articles were published in the journal *Asia* in the 39th volume in 1939. In the second article, Jung provided a very positive view of the Indian civilization. In his view, India was "more balanced psychologically than the West and hence less prone to the outbreaks of barbarism which at that time were only too evident" in Europe.[38] It is suggested that the concept of "self," as developed by Jung was largely based on the Upaniṣadic notion of *ātman*.[39] As Jung writes, "The East [India] teaches us another broader, more profound, and higher understanding - understanding through life."[40] According to Jung, the influence of Hindu literature is nothing new; such influences "may be found in the works of Meister Ekhart, Leibniz, Kant, Hegel, Schopenhauer, and E. von Hartmann."[41]

9.2.1 EMERSON AND THOREAU–TWO CELEBRATED AMERICAN SCHOLARS

Ralph Waldo Emerson (1803–1882) and Henry David Thoreau (1817–1862 CE) were two of the most celebrated American scholars of the nineteenth century. They proposed the philosophy of transcendentalism and thereby influenced generations of Americans and others worldwide. The worldviews of Emerson and Thoreau were influenced by the sacred books of the Hindus and other classics of India. Emerson was fond of reading the *Bhagavad-Gītā* and popularized the book among his friends by loaning his personal copy. This personal copy became quite worn in time due to its extensive use.[42] Emerson shared his appreciation for the Hindu philosophy in the following language: "This belief that the higher use of the material world is to furnish us types or pictures to express the thoughts of the mind, is carried to its extreme by the Hindoos [Hindus] . . . I think Hindoo [Hindu] books are the best gymnastics for the mind."[43] Thoreau also wrote quite approvingly about the Hindus: "the Hindoos [Hindus] . . . possess in a wonderful degree the faculty of contemplation," "their religious books describe the first inquisitive & contemplative access to God," or "their method is pure wisdom or contemplation."[44] Similar to the Hindu tradition, Thoreau promoted vegetarianism: "I believe that every man who has ever been earnest to preserve his higher or poetic faculties in the best condition has been particularly inclined to abstain from animal food, and from much food of any kind."[45]

Russell B. Goodman, in his analysis of Emerson's work, concludes that "Emerson's philosophy, from his college days onward, grew up together with his knowledge of and interest in

[37] Coward, 1984.
[38] Clarke, 1994, p. 62.
[39] Nicholson, Hanson, Stewart, 2002, p. 116.
[40] Jung, *Collected Works*, vol. 13, p. 7.
[41] Jung, *Letters*, vol. 1, p. 87.
[42] Horton and Edwards, 1952, p. 118; taken from Gangadharan, Sarma and Sarma, 2000, p. 311. For more information on Emerson, read Acharya, 2001.
[43] Emerson, 1904, vol. 8, p. 14.
[44] Scott, 2007.
[45] Henry David Thoueau, *Walden; or, Life in the Woods*, Dover, 1995, p. 139.

Hindu philosophical writings."[46] "His relationship with Hinduism, as with many other systems of thought, was transformative and respectful at once—reconstructive rather than deconstructive." At the youthful age of seventeen, Emerson wrote that during the ancient period in India "fair science pondered" and "sages mused."[47] Emerson wrote to John Chapman in May 1845 that he "very much want[ed]" the *Bhagavad-Gītā* to read the dialogs of Lord Kṛṣṇa and Arjuna.[48] In August 1845, Emerson went to the mountains of Vermont and read the *Viṣṇu-Purāṇa*. He commented that "[n]othing in theology is so subtle as this & the Bhagwat [perhaps *Bhāgvat-purāṇa*]."[49] Emerson's poem "Hamatreya" is based on a passage from the *Viṣṇu-Purāṇa*, and his essay "Immortality" is based on *Kathā Upaniṣad*. The beginning of his poem "Brahma" is also derived from *Kathā Upaniṣad* and the *Bhagavad-Gītā*, concludes Philip Goldberg, in his book *American Veda*.[50] "The borrowings were not aesthetic embellishments; they were central to Emerson's worldview."

Thoreau praised Hindu literature in his writing: "What extracts from the *Vedas* I have read fall on me like the light of a higher and purer luminary . . . It rises on me like the full moon after the stars have come out, wading through some far summer stratum of the sky."[51] A. K. B. Pillai, in his book *Transcendental Self: A Comparative Study of Thoreau and the Psycho-Philosophy of Hinduism and Buddhism*, evaluated the connections between the tenets of Hinduism and transcendentalism and concluded that "*Walden* is the closest to the Yogic system of all major American writings."[52] *Walden* is perhaps the most popular book of Thoreau.

"Whenever I have read any part of the Vedas, I have felt that some unearthly and unknown light illuminated me. In the great teaching of the Vedas, there is no touch of sectarianism. It is of all ages, climes and nationalities, and is the royal road for the attainment of the Great Knowledge," writes Thoreau.[53]

In *Walden*, a book written by Thoreau, he writes, "In the morning I bathe my intellect in the stupendous and cosmogonal philosophy of the Bhagvat-Geeta [*Bhagavad-Gītā*], since whose composition years of the gods have elapsed, and in comparison with which our modern world and its literature seem puny and trivial. . . I lay down the book and go to my well for water, and lo! there I meet the servant of the Brahmin, priest of Brahma [Brahmā] and Vishnu [Viṣṇu] and Indra, who still sits in his temple on the Ganges reading the *Vedas*, or dwells at the root of a tree with his crust and water jug. . . The pure Walden water is mingled with the sacred water of the Ganges."[54]

[46]Goodman, 1990.
[47]Goodman, 1990.
[48]Rusk, 1939, vol. III, p. 288; taken from Goodman, 1990.
[49]Goodman, 1990.
[50]Goldberg, 2010, p. 34.
[51]Thoreau, 1906, 2: 4.
[52]Pillai, 1985, p. 4, 88.
[53]Goldberg, 2010, p. 39.
[54]Taken from Scott, 2007

Thoreau and Emerson were the leaders of the American Transcendentalism movement of the nineteenth century.[55] They were drawn to Hindu texts since they received support to their own way of thinking in these texts. "There is little doubt that both Emerson and Thoreau shared an overwhelming and sustained enthusiasm for Vedic literature and Vedāntic philosophy and became their ardent votaries," concludes Sundararaman.[56] This support for Hinduism by Emerson and Thoreau has also been noticed by others. Raj Kumar Gupta shares on the enthusiasm and struggle Emerson and Thoreau dealt with: "[i]n their fervent and enthusiastic admiration for Hindu idealism and spiritualism, Emerson and Thoreau lost sight of the fact that they were contrasting American *practice* with Indian *theory*."[57] In his analysis, Gupta made the following judgment about Emerson and Thoreau: "Hindu ideas and ideals are used to bring out, explicitly or by implication, the failures and deficiencies of nineteenth century American life and thought, and in an attempt to fill the void and supply the deficiency. Thus, Hinduism represents not only ideals against which American values are judged and found wanting, but also a corrective and an antidote to those values."[58]

9.2.2 IMPACT ON PHYSICS

J. Robert Oppenheimer (1904–1967), an atomic physicist, director of the Manhattan Project, and the so-called father of the atomic bomb, not only read books of the ancient Hindus but even tried to learn the Sanskrit language to get the original intended meaning of them. While working with Ernest Lawrence in Berkeley, after he got his B.S. degree from Johns Hopkins University in 1933, he wrote to his brother Frank: "Lawrence is going to the Solvay conference on nuclei, and I shall have double chores in his absence . . . I have been reading the Bhagwad Gita [*Bhagavad-Gītā*] with Ryder and two other Sanskrists. It is very easy and quite marvelous. I have read it twice but not enough. . ."[59] In another letter to his brother Frank, he writes, "Benevolences starting with the precious Meghduta [a book by *Kālīdāsa*] and rather too learned *Vedas* . . ."[60] Within a year of study on the side, Oppenheimer became well versed in reading classics in Sanskrit on his own.[61] It is obvious that reading of the sacred books of the ancient Hindus shaped Oppenheimer's worldview and helped him in his researches in physics. Oppenheimer considered the *Bhagavad-Gītā* "the most beautiful philosophical song existing in any known tongue." He kept a well-worn copy of it conveniently on hand on the bookshelf closest to his desk and often gave the book (in translation) to friends as a present.[62]

[55] Christy, 1932; Riepe, 1970.

[56] Sundararaman, David Thoreau: The Walden-Ṛṣi, in the book by Gangadharan, Sarma, and Sarma, 2000, p. 311.

[57] Gupta, 1986, p. 81.

[58] Gupta, 1986, p. 81.

[59] Smith and Weiner, 1980, p. 162. The reference alludes to Arthur W. Rider (1877–1938), a professor of Sanskrit at the University of California, Berkeley.

[60] Smith and Weiner, 1980, p. 180.

[61] Nisbet, p. 136.

[62] Royal, 1969, 64.

In his condolence message upon the death of President Roosevelt, he asked people to have courage and continue with the task of the Manhattan Project at Los Alamos: "In the Hindu Scripture, in the Bhagwad Gita [*Bhagavad-Gītā*], it says, 'Man is a creature whose substance is faith. What his faith is, he is.' "[63] Most of his speech was taken from *Bhagavad-Gītā*. It is also well known that he started chanting the verses of *Bhagavad-Gītā* when he witnessed the test explosion of the atom bomb as director of the Manhattan Project. In the main verse, the aura of God is described as brighter than a thousand suns. It has led to a book an engaging and popular book on the atomic bomb with the same title.[64]

Brian Josephson (born 1940) received the Nobel Prize in physics in 1973 for his discovery of the Josephson quantum tunneling effect in superconductors. SQUID (Superconducting Quantum Interference Device), a magnetometer to measure low magnetic fields, is a device that is based on the Josephson effect. Josephson suffered from insomnia especially after he received Nobel Prize and was hooked on tranquilizers. In his attempt to protect his health, he started practicing transcendental meditation, as propagated by Maharshi Mahesh Yogi (1918–2008). These regular practices of meditation gave him "inner peace" and sound sleep. He is currently the director of the Mind-Matter Unification Project at the Cavendish Laboratory in England and a regular practitioner of yoga and meditation. His current researches mostly deal with the uncommon subjects of science: consciousness, the role of the observer and mind in quantum mechanics, analogies between quantum mechanics and Hindu mysticism, physics and spirituality, and yoga.[65] Josephson believes that scientists "can enhance their abilities through meditation."[66]

Erwin Schrödinger: the Guru of Dual Manifestations

Erwin Schrödinger, a Nobel Laureate in physics and one of the prominent architects of modern physics, writes on the connectedness of the various events in nature in his book, *My View of the World*: "Looking and thinking in that manner you may suddenly come to see, in a flash, the profound rightness of the basic conviction in Vedanta . . . Hence this life of yours is not merely a piece of the entire existence, but in a certain sense the whole. This, as we know, is what the Brahmins express in that sacred, mystical formula which is yet really so simple and so clear: *Tat tvam asi* (that is you). Or, again in such words as 'I am in the east and in the west, I am below and above, I am this whole world.'"[67] Vedanta is one of the six orthodox schools of Hindu philosophy. The word means "end of the Vedas," reflecting concepts that are documented in the Upaniṣads. *Tat tvam asi* is a Sanskrit term that mentioned in the *Chāndogya-Upaniṣad* (6: 8: 7) and connects self with the ultimate reality.

[63] Smith and Weiner, 1980, p. 288.

[64] Jungk, Robert, 1958. He was an Austrian writer who wrote extensively on nuclear weapons. He even ran for the presidency of Austria in 1992 for the Green Party.

[65] Josephson, 1987.

[66] Horgan, 1995

[67] Schrödinger, 1964, p. 21.

It is interesting to see similarities of Schrödinger's philosophy of life that he derived from the books of the Hindus and the quantum mechanical wavefunction he created to define the microscopic reality. Though anecdotal, it is interesting to note the similarities between quantum mechanical waves and wavefunctions with its all pervading nature, as defined by Schrödinger, with the all pervading (omnipresent) description of God. The wavefunction, though abstract, materializes itself in a region of space, in either its particle or its wave aspect, when squared (probability density). This is similar to the *Nirguṇa-svarūpa* (*amūrta*, without form) of God that manifests itself in human form (*saguṇa-svarūpa*) from time to time in different regions. It is important to mention that these analogies are not real connections. However, they do play an important role in the thinking of the inventors, as mentioned in Chapter 1. Schrödinger's worldview helped him in "hatching" scientific and mathematical ideas.

Schrödinger was influenced with the philosophy of Arthur Schopenhauer (1788–1860). Several other noted European intellectuals and philosophers were influenced by Schopenhauer, including Immanual Kant, Friedrich Nietzsche, Thomas Mann, Sigmund Freud, Albert Einstein, Carl Jung, and Leo Tolstoy.[68] Schopenhauer's philosophy was greatly influenced by the teachings of Hinduism and Buddhism. He suggested a life style of negating desires, similar to Pythagoras and the Hindu philosophy. His book, *The World as Will and Representation*, emphasizes the Upaniṣadic teaching: *Tat tvam asi* (that is you). Schopenhauer had a dog named Atman (meaning soul in Sanskrit). Though Schopenhauer was popular in Europe, his philosophy was not popular among physicists. Schrödinger realized that his views may not be appreciated by his physicist colleagues: "I know very well that most of my readers, despite Schopenhauer and the Upaniṣads, while perhaps admitting the validity of what is said here as a pleasing and appropriate metaphor, will withhold their agreement from any *literal* application of the proposition that all consciousness is essentially *one*."[69]

Observation is central to the growth of science. However, it provides a plurality of realities as the observation may differ from person to person, depending on the aspects these observers are interested in. This creates multifarious realities. This plurality does not get much attention by scientists. Schrödinger did consider the issue of plurality in life. "For philosophy, then, the real difficulty lies in the spatial and temporal multiplicity of observing and thinking individuals. . . the plurality that we perceive is only *an appearance; it is not real.* Vedantic philosophy, in which this is fundamental dogma, has sought to clarify it by a number of analogies, one of the most attractive being the many faceted crystal which, while showing hundreds of little pictures of what is in reality a single existent object, does not really multiply that object."[70] Schrödinger believed in the advocacy of the Hindu philosophy of oneness: "In all the world, there is no kind of framework within which we can find consciousness in the plural; this is simply something we construct because of the spatio-temporal plurality of individuals, but it is a false construction. . .

[68]Moore, 1992, p. 112
[69]Schrödinger, 1964, p. 29.
[70]Schrödinger, 1964, p. 18.

The only solution to this conflict, in so far as any is available to us at all, is in the ancient wisdom of the Upanishads."[71]

In his essay on *The Diamond Cutter*, Schrödinger writes: "The ego is only an aggregate of countless illusions, a phantom shell, a bubble sure to break. It is *Karma*. Acts and thoughts are forces integrating themselves into material and mental phenomena—into what we call objective and subjective appearances . . . The universe is the integration of acts and thoughts. Even swords and things of metal are manifestations of spirit. There is no birth and death but the birth and death of Karma in some form or condition."[72] These statements led Walter Moore, who has written a book on the life of Schrödinger, to infer that Schrödinger "thought deeply about the teachings of Hindu Scriptures, reworked them into his own words, and ultimately came to believe in them."[73] Also, his worldview helped him in creating the wave-mechanics which is counterintuitive to Newtonian thinking: "Perhaps these thoughts recurred to Erwin when he made his great discovery of wave mechanics and found the reality of physics in wave motions, and also later when he found that this reality was part of an underlying unity of mind."[74]

The philosophy of multiple representations of truth, thin boundaries between abstract thoughts (*amūrtā*), actual realities (*mūrta*) and universal consciousness played an important role in the science that Erwin Schrödinger created. Moore concludes, "Vedanta and gnosticism are beliefs likely to appeal to a mathematical physicist, a brilliant only child, tempted on occasion by intellectual pride. Such factors may help to explain why Schrödinger became a believer in Vedanta, but they do not detract from the importance of his belief as a foundation for his life and work. It would be simplistic to suggest that there is a direct causal link between his religious beliefs and his discoveries in theoretical physics, yet the unity and continuity of Vedanta are reflected in the unity and continuity of wave mechanics."[75] In the view of Moore, "Schrödinger and Heisenberg and their followers created a universe based on superimposed inseparable waves of probability amplitudes. This view would be entirely consistent with the Vedantic concept of the All in One."[76]

* * * * * * * *

Just think of the global impact of the efforts of the ancient Hindus for knowing the ultimate truth. They created disciplines that were for the welfare of all humanity. Hinduism has transcended the boundaries of what is called 'religion' in the West and has permeated the whole world. Yoga is practiced by devoted Christians, Muslims, and atheists, without discrimination. It is currently prescribed in many hospitals in the Western world for pain management and cures of various diseases. The United Nations have declared June 21 as the "yoga day" which is being celebrated all over the world.

[71] Schrödinger, 1964, p. 31.
[72] taken from Moore, 1992, p. 114.
[73] Moore, 1992, p. 113.
[74] Moore, 1992, p. 114.
[75] Moore, 1992, p. 173.
[76] Moore, 1992, p. 173.

Think of the kids that are schooled in Africa, Asia, or Europe. Everywhere in the civilized world, kids learn to count even before they learn the alphabet of their own language. In different languages, the numbers may be named differently; but, they all use the same (base 10) mathematical principles (known as the decimal system) in counting and in arithmetic operations that were invented long ago by the ancient Hindus. These mathematical methods are so effective and important, most kids learn them without ever inquiring about the alternatives. "The Indian system of counting has been the most successful intellectual innovation ever made on our planet. It has spread and been adopted almost universally, far more extensively even than the letters of the Phoenician alphabet which we now employ. It constitutes the nearest thing we have to a universal language,"[77] writes John D. Barrow in his book, *Pi in the Sky: Counting, Thinking, and Being.*

In dealing with numbers, the size of the atom conceived by the ancient Hindus comes close to 10^{-10}m. For the age of the universe, they came up with a number that is of the order of a billion years. Both of these numbers were considered mere speculations and ridiculed by foreign scientists during the early medieval period. However, these numbers are in tune with the modern science now. Is this a coincidence? How can one reach such striking conclusions that were considered ridiculous by scientists just a century ago? For the Hindus, this information was valuable enough to be preserved in their most sacred books. How did they reach to these numbers? It is not an easy question to answer. The answer is perhaps in the role of the mind in the meditative state to learn the mysteries of nature that are not otherwise accessible through the faculties of the five senses we human beings are endowed with. This role of mind or consciousness is what Noble Laureate Brian Josephson is investigating these days at the University of Cambridge, London and has attracted the minds of Robert Oppenheimer, Erwin Schrödinger, and others.

The social attitudes in India toward accepting new scientific theories were in contrast to those prevailing in Europe. Unlike Galileo and Copernicus, Āryabhaṭa never faced any hostile sentiments or violent reaction from the priestly class or the possibility of any kind of persecution from the king when he assigned motion to the earth. On the contrary, invention and promotion of such ideas led a person to the greatest honor on earth while alive, and *mokṣa* - ultimate goal of liberation - after death, as suggested by Āryabhaṭa I. Thus, becoming an astronomer, a medical doctor, or a musician was as honorific profession as becoming a priest or swami. This is perhaps the reason why Caraka and Suśruta were so deeply involved in the medical research and practice and Āryabhaṭa in astronomy and mathematics. Ultimately, they were working toward achieving *mokṣa*, liberation of their soul.

Multicultural Cooperation in Science

Many intellectual centers in our history used multicultural approaches to learning. Damascus and Baghdad became leading centers of learning when they started attracting Greek, Indian,

[77]Barrow, 1992, p. 92.

Persian, and Egyptian scholars. Similarly, Spain became a center of learning from the eighth century to the eleventh century, because it attracted scholars from all over the world.[78] However, these centers eventually declined when the scholars were judged on ill-conceived grounds (race, gender, religion, or culture), rather than on the basis of merit. Such declines occurred in Spain, Egypt, the Middle East, and elsewhere in Europe.

Modern science is international in character. Scientific ideas are studied and developed in many languages from all over the world. Thus, science education with a multicultural approach can be a broadening and humanizing experience. Multicultural approach teaches us how human beings are alike in the pursuit of knowledge. It inspires us to learn from each other and appreciate each other. It allows us to function as a cohesive society. However, so far, science education has not utilized this opportunity to its fullest potential.

Preserving knowledge is a process in which all generations must participate or become susceptible to lose some valuable knowledge. Museums are created to slow down or stop this tide of entropy. However, they mostly secure artifacts and have limitations. Perhaps, libraries are the places where knowledge can be documented and preserved. However, these libraries are prone to destruction as we know from the history of Alexandria, Nālandā, and Taxila. Therefore, dissemination of knowledge to the next generation is a duty that each generation must involve in. This is called *guru-ṛṇa* (debt to the guru) that each disciple owes to his or her own teacher to continue the tradition of knowledge, to carry forward the torch (light) of knowledge to the next generation.

No culture or civilization has prospered to great heights without knowing and preserving their historic and existing knowledge base. This book is written to preserve the knowledge of the ancient Hindus and to recognize their rightful place in world history. Of course, it is only a small step; much more effort needs to be exercised on a continual basis to research, disseminate and preserve this ancient knowledge of the Hindus from the earliest period of the Indus-Sarasvati civilization to the end of the last millennium when this knowledge was deliberately suppressed in the pursuit of colonial objectives. The twenty-first century is the harbinger of free, open and honest pursuit of scientific knowledge in all its varied dimensions unfettered by dogma or national or cultural chauvinism.

[78] Salem and Kumar, 1991.

APPENDIX A

Timeline of the Hindu Manuscripts

A lack of well-accepted chronology is the weakest point of the Hindu manuscripts. There are several reasons for it: first, the Hindu culture practiced orality to preserve their ideas unlike the Egyptian (papyrus) and Babylonians (clay tablets). It is difficult to find archaeological records in a oral culture. Second, the Hindu culture basically followed in the *same tradition* from the earliest periods. As discussed in Chapter 5, events are essential to define time not only in the physical sense but in cultural sense too. Events of atrocities, change in value system, change in prosperity, etc. are the landmarks of a culture. However, in the case of the Hindus, a similar pattern of prosperity and cultural traits continued for so long, the culture paid little or no attention to define events and time frame.

We know that the *Ṛgveda* is the earliest book of the Hindu literature, and is certainly one of the earliest known manuscripts in human history. The ancient Hindus did not erect monuments to document history, until the period of Emperor Aśoka who reigned in India between 268 to 232 BCE. The ancient Hindus did not even mention the names of the authors in their books. Some Western historians look for a "hard-copy" or "hard-evidence" to accept a particular date for the antiquity of the Hindu manuscripts. Due to the prevalent oral tradition, such evidences are not possible with the ancient Hindus. Jessica Frazier, a Lecturer in Religious Studies at the University of Kent and a Fellow of the Oxford Center for Hindu studies, in her quest to define history, historiography, and timeline of various Hindu events, shares her experience in the following words: "The search for an Indian historiography is like the work of a geologist who must sift innumerable layers of compacted material, or perhaps better, a zoologist who can never stop seeking new species of history, and never assume that any given specimen will remain long in a single place or form." She criticizes too much reliance on the accounts of travelers such as Herodotus and others who visited India and wrote their accounts. In her view, "[t]he ideal Western historian has classically been seen as an itinerant Godlike traveller." These travellers had a limited exposure of the culture and wrote their view from their bird's eye view. Even the astronomical observations of the heavenly bodies which should provide a linear scale of time fails because of "multiple parallel chronologies." She concludes that, "'while India's Hindu past was ever-present in divine reality and recursive myth, it was nevertheless unreachable by accepted historiographical methods."[1]

[1] Frazier, 2009.

During the colonial period, scholars in the West worked hard to define Hindu religion and culture in their efforts to Christianize India. At that time in Europe, it was a prevalent belief among Christians that the world was created around 4004 BCE on one Friday afternoon. This was based on the genealogy presented in the book of *Genesis*, in the *Bible*, as concluded by Archbishop James Ussher (1581–1656 CE), who was also the Vice-Chancellor of the Trinity College in Dublin. In 1642, Dr. John Lightfoot (1602–1675 CE), then Vice Chancellor at the Cambridge University improved the calculation further and made it more precise to be 9 AM, October 23, 4004 BCE.[2] These calculations were generally agreed on by the Western scholars at the beginning of the twentieth century. As a result, the Western scholars assigned dates to *Vedas* and *upaniṣad* that made sense to the misconstrued date of 4004 BCE. Once these dates were arbitrarily defined, they have not been modified with current research.

Al-Bīrūnī (973–1050 CE) noticed the problem of historiography in assigning actual dates to various events in the eleventh century and criticized the ancient Hindus: "Unfortunately the Hindus do not pay much attention to the historical order of things, they are very careless in relating the chronological succession of their kings."[3]

Carl Jung (1875–1961 CE), a psychoanalyst and philosopher, aptly answered such criticisms of the absence of a chronological history in India, by attributing it to the antiquity of the Hindu civilization. "After all why should there be recorded history [in India]? In a country like India one does not really miss it. All her native greatness is in any case anonymous and impersonal, like the greatness of Babylon and Egypt. History makes sense in European countries where, in a relatively recent, barbarous, and inhistorical past, things began to shape up . . . But in India there seems to be nothing that has not lived a hundred thousand times before."[4] Even the ancient writers such as Caraka (around 600 BCE), Suśruta (around 1000 BCE), and Ārybhaṭa I (5th century) do not represent the first efforts for most theories in their books; they all claim to be merely reproducing old ideas in new form.

The first book in the modern era on the Hindu sciences was published by Brajendranath Seal in 1915.[5] Seal opted to deal with the absence of a proper chronology right in the first paragraph of his Foreword. He basically assigned an umbrella period of 500 BCE to 500 CE to all Hindu manuscripts considered in his book. Bose *et al* in 1971 and Chattopadhyaya in 1986, two prominent books on the history of Hindu science, encountered the same issue. This situation has not improved much in the last century; we still cannot assign accurate dates for ancient manuscripts.

The issue of antiquity can only be solved either from the artifacts of the ancient Hindus or from the astronomical maps provided in their oral literature. Previously, the maps provided in the Hindu oral literature were ignored by the scholars as they defined the period of *Ṛgveda* to be more ancient than 1500 BCE, an arbitrary date assigned by Max Müller. Since the recent

[2]White, 1897, p. 9.
[3]Sachau, 1964, vol. 2, p. 10.
[4]Jung, *Collected Works*, vol. 10, p. 517.
[5]Seal, 1915

excavations unearthed artifacts that are some 10,000 years old, a need for the revision of history is warranted.[6] A massive effort is needed to resolve this issue.

Though the dates are not certain, there is a general acceptance about the chronological order of these books and periods can be assigned to these books, as listed in Table A.1. The content of my book will not change much with corrections in this table. Most conclusions will remain valid even with a different chronology.

Table A.1: The Antiquity of Hindu Books

Manuscript	Date of Period
Āryabhaṭīya	5th century CE
Bakhshālī Manuscript	Seventh Century CE
Brāhmaṇas	Between 2000 - 1500 BCE
Mahābhārata Battle	Between 2449 BCE to 1424 BCE
Pingala's Chandah Sutra	Between 480 - 410 BCE
Purāṇa	Between the Christian era and 11th century
Manu-Saṁhitā ,	Between 500 - 100 BCE
Caraka-Saṁhitā	Around 600 BCE
Suśruta-Saṁhitā	Around 1000 BCE
Śulbasūtra	Between 1700 BCE to 1000 BCE
Upanishads	800 to the beginning of the Christian era
Vaiśesika-Sūtra	Between 900 - 600 BCE
Vedas	Around 1500 BCE or earlier

[6]Kenoyar, 1998.

References

[1] *Atharvaveda: The Hymns of the Atharvaveda*, Ralph T. H. Griffith, Chowkhamba Sanskrit Series, Varanasi, 1968.

[2] *Bhagavad-Gītā*, S. Radhakrishnan, Unwin Paperbacks, London, 1948.

[3] *Caraka Saṁhitā: The Caraka Saṁhitā*, Satyanārāyaṇa Śāstrī, Chowkhamba Sanskrit Series, Varanasi, 1992, 2 volumes, in Hindi.

[4] *Kauṭilya's Arthaśāstra: Kauṭilya's Arthaśāstra*, Shamasastry, R. Mysore Printing and Publishing House, Mysore, 1960. 4th ed., (first published in 1915).

[5] *Purāṇa*, Translator: Pandit Shreeram Ji Sharma, Sanskrit Sansthan, Bareiley, 1988. In Hindi and Sanskrit. All eighteen *Purāṇas* have been translated by the same author and published from the same publications.

[6] *Ṛgveda: The Hymns of the Ṛgveda*, Ralph T. H. Griffith, Chowkhamba Sanskrit Series, Varanasi, 1963.

[7] *Suśruta-Saṁhitā: The Suśruta-Saṁhitā*, Bhishagratna, Kaviraj Kunjalal (Translator), Chaukhambha Sanskrit Series, Varanasi, 1963, 3 volumes.

[8] *Upaniṣads*, Translator: Patrick Olivelle, Oxford University Press, Madras, 1996.

[9] *Upaniṣads*: *The Thirteen Principal Upanishads*, Translator: Robert Ernst Hume, Oxford University Press, Madras, 1965.

[10] *Upaniṣads*: *108 Upanishads*, Translator: Pandit Shreeram Ji Sharma, Sanskrit Sansthan, Bareiley, 1990. In Hindi and Sanskrit.

[11] *Vedas*, Translator: Pandit Shreeram Ji Sharma, Sanskrit Sansthan, Bareiley, 1988. In Hindi and Sanskrit. All four *Vedas* have been translated by the same author and published from the same publications. DOI: 10.1093/acprof:oso/9780195658712.003.0022.

[12] *Viṣṇu Purāṇa*, Wilson, H. H. (Translator), Punthi Pustak, Calcutta, 1961.

[13] *Yajurveda (Vājasaneyisamhitā): The Texts of the White Yajurveda*, Ralph T. H. Griffith, Chowkhamba Sanskrit Series, Varanasi, 1976.

[14] Achar, B. N. Narahari, Āryabhaṭa and the tables of Rsines, *Indian Journal of History of Science*, 37(2), 95–99, 2002.

[15] Acharya, Shanta, *The Influence of Indian Thought on Ralph Waldo Emerson*, The Edwin Mellen Press, Lewiston, 2001.

[16] Achaya, K. T., Alcoholic fermentation and its products in ancient India, *Indian Journal of History of Science*, 26(2), 123–129, 1991.

[17] Ackernecht, Erwin Heinz, *A Short History of Medicine*, Johns Hopkins University Press, Maryland, 1982.

[18] Aelianus, Claudius, *On the Characteristics of Animals*, A. F. Scholfield (Translator), Harvard University Press, Cambridge, 1958.

[19] Agarwal, D. P. *Ancient Metal Technology and Archaeology of South Asia: A Pan-Asian Perspective*, Aryan Books International, New Delhi, 2000.

[20] Al-Kindī, *The Medical Formulary or Aqrābādhīn of Al-Kindī*, Martin Levey (Translator), The University of Wisconsin Press, Madison, 1966.

[21] Alter, Joseph S., Heaps of health, metaphysical fitness, *Current Anthropology*, 40, S43–S66, 1999. DOI: 10.1086/200060.

[22] Ang, Gina C., History of skin transplantation, *Clinics in Dermatology*, 23, 320–324, 2005. DOI: 10.1016/j.clindermatol.2004.07.013.

[23] Aristotle, *Aristotle Minor Works*, W. S. Hett (Translator), Harvard University Press, Cambridge, 1963 (first published 1936).

[24] Arenson, Karen W., When Strings are Attached, Quirky Gifts can Limit Universities, *The New York Times*, April 13, 2008.

[25] Baber, Zaheer, *The Science of Empire: Scientific Knowledge, Civilization, and Colonial Rule in India*, State University of New York, Albany, 1996.

[26] Bacon, Roger, *The Opus Majus of Roger Bacon*, Translator: Robert Belle Burke, University of Pennsylvania Press, Philadelphia, 1928. (Originally published with Oxford University Press). DOI: 10.1017/cbo9780511709678.

[27] Bag, A. K., *Mathematics in Ancient and Medieval India*, Chaukhambha Orientalia, Delhi, 1979.

[28] Bagchi, P. C., *India and China*, Hind-Kitab, Bombay, 1950.

[29] Bailey, Cyril, *The Greek Atomists and Epicurus*, Clarendon Press, Oxford, 1928.

[30] Balasubramaniam, R., V. N. Prabhakar, and Manish Shankar, On technical analysis of cannon shot crater on Delhi iron pillar, *Indian Journal of History of Science*, 44(1), 29–46, 2009.

[31] Balasubramaniam, R. and A. V. Ramesh Kumar, Corrosion resistance of the Dhar iron pillar, *Corrosion Science*, 45, 2451–2465, 2003. DOI: 10.1016/s0010-938x(03)00074-x.

[32] Balasubramaniam, B., A new study of Dhar iron pillar, *Indian Journal of History of Science*, 37(2), 115–151, 2002.

[33] Balasubramaniam, B., New insights on the 1,600 year-old corrosion resistant Delhi iron pillar, *Indian Journal of History of Science*, 36(1–2), 1–49, 2001.

[34] Balslev, A. N., *A Study of Time in Indian Philosophy*, Otto Harrassowitz, Wiesbaden, 1983.

[35] Barua, Beni Madhab, *Aśoka and His Inscriptions*, New Age Publishers, Calcutta, 1946.

[36] Barrow John D., *Pi in the Sky: Counting, Thinking, and Being*, Oxford University Press, New York, 1992.

[37] Bays, G., *The Voice of Buddha*, Dharma Publishing, California, 1983 (original title, *Lalitvistara*).

[38] Beal, Samuel, *Si-Yu-Ki Buddhist Records of the Western World*, Paragon Book Reprint Corp., New York, 1968. DOI: 10.4324/9781315011783.

[39] Bernal, J. D., *Science in History*, The MIT Press, Cambridge, 1971.

[40] Bernstein, Richard, *Ultimate Journey*, Alfred A. Knopf, New York, 2001.

[41] Bhagvat, R. N., Knowledge of the metals in ancient India, *Journal of Chemical Education*, 10, 659–666, 1933. DOI: 10.1021/ed010p659.

[42] Bhardwaj, H. C., *Aspects of Ancient Indian Technology: A Research Based on Scientific Methods*, Motilal Banarsidas, Delhi, 1979.

[43] Bigwood, J. M. Ctesias, His royal patrons and Indian swords, *Journal of Hellenic Studies*, 115, 135–140, 1995. DOI: 10.2307/631649.

[44] Billard, Roger, Aryabhaṭa and Indian astronomy, *Indian Journal History of Science*, 12(2), 207–224, 1977,

[45] Biswas, Arun Kumar, and Sulekha Biswas, *Minerals and Metals in Ancient India*, D. K. Printworld, New Delhi, 1996, 3 volumes.

[46] Biswas, Arun Kumar, and Sulekha Biswas, *Minerals and Metals in Pre-Modern India*, D. K. Printworld, New Delhi, 2001.

[47] Boncompagni, B., Ed., *Scritti di Leonardo Pisano*, Rome, 1857–1862, 2 volumes (in Italian).

[48] Bose, D. M., S. N. Sen, and B. V. Subbarayappa, *A Concise History of Science in India*, Indian National Science Academy, New Delhi, 1971.

[49] Bose, S. K., The atomic hypothesis, *Current Science*, 108, 998–1002, March 10, 2015.

[50] Brain, David J., The early history of rhinoplasty, *Facial Plastic Surgery*, 9(2), 91–88, April 1993. DOI: 10.1055/s-2008-1064600.

[51] Brennard, W., *Hindu Astronomy*, Caxton Publications, Delhi, 1988.

[52] Brier, Bob, Napoleon in Egypt, *Archaeology*, 52(3), 44–53, May/June 1999.

[53] Bronkhorst, Johannes, A note on zero and the numberican place-value system in ancient India, *Asiatische Studien Études Asiatiques*, 48(4) 1039–1042, 1994.

[54] Brown, Ronald A. and Alok Kumar, A new perspective on eratosthenes' measurement of earth, *The Physics Teacher*, 49, 445–447, October 2011. DOI: 10.1119/1.3639158.

[55] Bubnov, Nikolai Michajlovic, *Gerberti, Opera Mathematica*, Berlin, 1899 (in Latin).

[56] Burgess, J., Notes on Hindu astronomy and the history of our knowledge of it, *Journal of Royal Asiatic Society*, 717–761, 1893. DOI: 10.1017/s0035869x00022553.

[57] Burnett, Charles, *The Introduction of Arabic Learning into England*, The British Library, London, 1997.

[58] Burnett, Charles, The semantics of Indian numerals in Arabic, Greek and Latin, *Journal of Indian Philosophy*, 34, 15–30, 2006. DOI: 10.1007/s10781-005-8153-z.

[59] Cajori, Florian, *A History of Mathematics*, Chelsea Publishing Company, New York, 1980.

[60] Calinger, Ronald, *A Contextual History of Mathematics*, Prentice Hall, NJ, 1999 (first published in 1893).

[61] Capra, Fritjof, *The Tao of Physics*, Bantam New Age Books, New York, 1980. DOI: 10.1016/b978-0-08-028127-8.50009-x.

[62] Carpue, J. C., *An Account of Two Successful Operations for Restoring Lost Nose, from the Integuments of the Forehead*, Longman, London, 1816. DOI: 10.1097/00006534-198208000-00030.

[63] Casson, Lionel, *The Periplus Maris Erythraei*, Princeton University Press, 1989. DOI: 10.1017/s0009838800031840.

[64] Chabás, José, The diffusion of the Alfonsine tables: The case of the tabulae resolutae, *Perspectives on Science*, 10(2), 168–178, 2002.

[65] Chapple, Christopher Key and Mary Evelyn Tucker, Eds., *Hinduism and Ecology: The Intersection of Earth, Sky and Water*, Harvard Divinity School, Cambridge, 2000.

[66] Chaudhary, Anand and Neetu Singh, Herbo mineral formulations (rasaoushadhies) of Ayurveda: An amazing inheritance of Ayurvedic pharmaceutics, *Ancient Science of Life*, 30(1), 18–26, 2010.

[67] Chattopadhyaya, D., *History of Science and Technology in Ancient India*, Firma KLM Pvt. Ltd., Calcutta, 1986.

[68] Chikara, Sasaki, Mitsuo Sugiura, and Joseph W. Dauben, Eds., *The Intersection of History and Mathematics*, Birkhäuser Verlag, Basel, 1994.

[69] Christy, Arthur, *The Orient in American Transcedentalism: A Study of Emerson, Thoreau and Alcott*, Columbia University Press, New York, 1932.

[70] Clark, W. E., *The Āryabhaṭīya of Ārybhaṭa I*, The University of Chicago Press, Illinois, 1930.

[71] Clark, W. E., *Indian Studies in the Honor of Charles Rockwell Lanman*, Harvard University Press, Cambridge, 1929.

[72] Clarke, J. J., *Jung and Eastern Thought: A Dialogue with the Orient*, Routeledge, New York, 1994. DOI: 10.4324/9780203138533.

[73] Colebrooke, Henry Thomas (Translator), *Algebra with Arithmetic and Mensuration from the Sanscrit of Brahemegupta and Bhascara*, John Murray, London, 1817.

[74] Conboy, Lisa. A., Ingrid Edshteyn, and Hilary Garivsaltis, Ayurveda and panchakarma: Measuring the effects of a holistic health intervention, *The Scientific World Journal*, 9, 272–280, 2009. DOI: 10.1100/tsw.2009.35.

[75] Cooke, Roger, *The History of Mathematics*, John Wiley & Sons, Inc., New York, 1997. DOI: 10.1002/9781118033098.

[76] Copernicus, Nicholas, *On the Revolutions*, E. Rosen (translator), The Johns Hopkins University Press, Maryland, 1978.

[77] Coppa, A., L. Bondioli, A. Cucina, D. W. Frayer, C. Jarrige, J. -F. Jarrige, G. Quivron, M. Rossi, M. Vidale, and R. Macchiarelli, Palaeontology: Early neolithic tradition of dentistry, *Nature*, 440, 755–6, April 6, 2006.

[78] Coulter, H. David, *Anatomy of Hatha Yoga*, Body and Breath, Pennsylvania, 2001.

[79] Coward, H., Jung and Hinduism, *The Scottish Journal of Religions*, 5, 65–68, 1984, also read *Jung and Eastern Thought*, State University of New York Press, 1985.

[80] Cowen, Virginia, Functional fitness improvements after a worksite-based yoga initiative, *Journal of Bodywork and Movement Therapies*, 14, 50–54, 2010. DOI: 10.1016/j.jbmt.2009.02.006.

[81] Crossley, J. N. and A. S. Henry, Thus Spake al-Khwarizmi, *Historia Mathematica*, 17, 103–131, 1990.

[82] Danucalov, Marcello A., Roberto S. Simões, Elisa H. Kozasa, and José Roberto Leite, Cardiorespiratory and metabolic changes during yoga sessions: The effects of respiratory exercises and meditation practices, *Applied Psychophysiology and Biofeedback*, 32(2), 77–81, 2008. DOI: 10.1007/s10484-008-9053-2.

[83] Darlington, Oscar, G., Gerbert, The teacher, *The American Historical Review*, 52(3), 456–476, 1947.

[84] Dasgupta, Subrata, Jagadis Bose, Augustus Waller and the discovery of "vegetable electricity," *Notes and Records of the Royal Society of London*, 52(2), 307–322, 1998. DOI: 10.1098/rsnr.1998.0052.

[85] Dasgupta, S. N., *History of Indian Philosophy*, Cambridge University Press, Cambridge, 1922–1955, 5 volumes.

[86] Dasgupta, A., *Goethe and Tagore*, South Asia Institute, Heidelberg, 1973.

[87] Dasgupta, S. N., *History of Indian Philosophy*, Cambridge University Press, Cambridge, 1922–1955, 5 volumes.

[88] Dasgupta, Surendra, *A History of Indian Philosophy*, Cambridge University Press, Cambridge, 1963.

[89] Datta, B. B., *The Science of Sulba*, University of Calcutta, Calcutta, 1932.

[90] Datta, Bibhutibhusan, On mula, the Hindu term for "root," *The American Mathematical Monthly*, 34(8), 420–423, October, 1927. DOI: 10.2307/2299170.

[91] Datta, B. B., Early literary evidence of the use of zero in India, *American Mathematical Monthly*, 33, 449–454, 1926. DOI: 10.2307/2299608.

[92] Datta, Bibhutibhusan and Avadhesh Narayan Singh, *History of Hindu Mathematics*, Asia Publishing House, New York, 2 volumes, 1961 (first published in 1938).

[93] Deshpande, Vijaya Jayant, Glimpses of ayurveda in medieval Chinese medicine, *Indian Journal of History of Science*, 43(2), 137–161, 2008.

[94] Deshpande, Vijaya J., Allusions to ancient Indian mathematical sciences in an early eighth century Chinese compilation by Gautama Siddha, *Indian Journal of History of Science*, 50(2), 215–226, 2015. DOI: 10.16943/ijhs/2015/v50i2/48237.

[95] Dikshit, Moreshwar G., *History of Indian Glass*, University of Bombay, Bombay, 1969.

[96] Dollemore, Doug, Mark Giuliucci, Jennifer Haigh, and Sid Kirchheimer, Jean Callahan, *New Choices in Natural Healing*, Rodale Press, Inc., Emmaus, Pennsylvania, 1995.

[97] Durant, Will, *Our Oriental Heritage*, Simon and Schuster, New York, 1954.

[98] Dwivedi, O. P., Vedic heritage for environmental stewardship, *Worldviews: Environment, Culture, Religion*, 1, 25–36, 1997.

[99] Einstein, Albert, On Prayer; Purpose in Nature; Meaning of Life; the Soul; A Personal God, *New York Times Magazine*, 1–4, November 9, 1930.

[100] Eliade, Mircea, *Yoga: Immortality and Freedom*, Translated by Willard R. Trask, Princeton University Press, NJ, 1969.

[101] Eliot, Charles, *Hinduism and Buddhism*, Routledge and Kegan Paul Ltd., London, 1954, 3 volumes.

[102] Emerson, Waldo, *The Complete Works of Ralph Waldo Emerson*, Houghton, Miffin and Company, Cambridge, 1904, 10 volumes.

[103] Faraday, M., An analysis of wootz or Indian steel, *Quarterly Journal of Science, Literature, and the Arts*, 7, 319–330, 1819.

[104] Farrington, B., *Science in Antiquity*, Oxford University Press, London, 1969 (first published in 1936).

[105] Feuerstein, Georg, *The Yoga-sutra of Patañjali*, Inner Traditions International, Rochester, Vermont, 1989.

[106] Fiala, Nathan, Meeting the demand: An estimation of potential future greenhouse gas emissions from meat production, *Ecological Economics*, 67, 412–419, 2008. DOI: 10.1016/j.ecolecon.2007.12.021.

[107] Figiel, L. S., *On Damascus Steel*, Atlantas Arts Press, Atlanta, 1991.

[108] Filliozat, J., *The Classical Doctrine of Indian Medicine: Its Origins and its Greek Parallels*, (Translator) Dev Raj Chanana, Munshiram Manoharlal, Delhi, 1964.

[109] Filliozat, Pierre-Sylvain, Making something out of nothing, *UNESCO Courier*, 30–34, November 1993.

[110] Findly, Ellison B., Jahāngīr's vow of non-violence, *Journal of the American Oriental Society*, 107(2), 245–256, 1987. DOI: 10.2307/602833.

[111] Folkerts, Menso, Early texts on Hindu-Arabic calculation, *Science in Context*, 14(1/2), 13–38, 2001. DOI: 10.1017/s0269889701000023.

[112] Frazier, Jessica, History and historiography in Hinduism, *The Journal of Hindu Studies*, 2, 1–16, 2009. DOI: 10.1093/jhs/hip009.

[113] Gangadharan, N., S. A. S. Sarma, and S. S. R. Sarma, *Studies on Indian Culture, Science, and Literature*, Sree Sarada Education Society, Chennai, 2000.

[114] Garratt, G. T., *Legacy of India*, Clarendon Press, Oxford, 1938.

[115] Gazalé, Midhat, *Number: From Ahmes to Cantor*, Princeton University Press, NJ, 2000.

[116] Ghosh, Pranabendra Nath, *Johann Gottfried Herder's Image of India*, Visva-Bharati Research Publication, Calcutta, 1990.

[117] Gies, Joseph and Frances Gies, *Leonardo of Pisa and the New Mathematics of the Middle Ages*, Thomas Y. Crowell Company, New York, 1969.

[118] Gilbert, Christopher, Yoga and breathing, *Journal of Bodywork and Movement Therapies*, 3(1), 44–54, 1999a. DOI: 10.1016/s1360-8592(99)80042-4.

[119] Gilbert, Christopher, Breathing and the cardiovascular system, *Journal of Bodywork and Movement Therapies*, 3(4), 215–224, 1999b. DOI: 10.1016/s1360-8592(99)80006-0.

[120] Gingerich, Owen, Ed., *The Eye of Heaven: Ptolemy, Copernicus, and Kepler*, American Institute of Physics, New York, 1993.

[121] Gohlman, William E., *The Life of Ibn Sina*, State University of New York Press, Albany, 1974.

[122] Goldberg, Philip, *American Veda*, Three Rivers Press, New York, 2010.

[123] Goldstein, B. R., Astronomy and astrology in the works of Abraham Ibn Ezra, *Arabic Sciences and Philosophy*, 6(1), 9–21, 1996. DOI: 10.1017/s0957423900002101.

[124] Goldstein, B. R., *Ibn al-Muthannâ's Commentary on the Astronomical Tables of al-Khwârizmī*, Yale University, New Haven, 1967.

[125] Goodman, Russell B., East-west philosophy in 19th-century America: Emerson and Hinduism, *Journal of the History of Ideas*, 51(4), 635–645, 1990. DOI: 10.2307/2709649.

[126] Goonatilake, S., *Aborted Discovery: Science and Creativity in the Third World*, Zed Books, London, 1984.

[127] Goonatilake, S., The voyages of discovery and the loss and rediscovery of "others" knowledge, *Impact of Science on Society*, 167, 241–264, 1992.

[128] Gorini, Catherine A., Ed., *Geometry at Work*, The Mathematical Association of America, 2000.

[129] Gosling, David L., *Religion and Ecology in India and Southeast Asia*, Routledge, New York, 2001.

[130] Gregorios, Paulos Mar, Ed., *Neoplatonism and Indian Philosophy*, State University of New York Press, 2002.

[131] Gupta, R. C., Indian astronomy in China during ancient times, *Vishveshvaranand Indological Journal*, India, 9, 266–276, 1981.

[132] Gupta, R. C., Spread and triumph of Indian numerals, *Indian Journal of the History of Science*, 18, 23–38, 1983.

[133] Gupta, R. C., Who Invented the zero?, *Gaṇita Bhāratī*, 17(1–4), 45–61, 1995.

[134] Gupta, Raj Kumar, *Great Encounter: A Study of India-American Literature and Cultural Relations*, Abhinav Publications, New Delhi, 1986.

[135] Halbfass, W., *India in Europe*, State University of New York, Albany, 1988.

[136] Hammett, Frederik S., The ideas of the ancient Hindus concerning man, *ISIS*, 28(1), 57–72, 1938. DOI: 10.1086/347304.

[137] Harding, Sandra, *Whose Science? Whose Knowledge?*, Cornell University Press, Ithaca, 1991. DOI: 10.7591/9781501712951.

[138] Harding, Sandra, Is science multicultural? Challenges, resources, opportunity, uncertainties, *Configuration*, 2, 301–330, 1994. DOI: 10.1353/con.1994.0019.

[139] Hayasi, T., Āryabhaṭa's rule and table for sine-differences, *Historia Mathematica*, 24, 396–406, 1997. DOI: 10.1006/hmat.1997.2160.

[140] Hayashi, T, Takanori Kusuba, and Michio Yano, Indian values for π from Āryabhaṭa's value, *Historia Scientiarum*, 37, 1–16, 1989.

[141] Hayasi, T., *The Bakhshālī Manuscript: An Ancient Indian Mathematical Treatise*, Egbert Forsten, Groningen, 1995.

[142] Herschel, Sir John, Sir John Herschel on Hindu mathematics, *The Monist*, 25(2), 297–300, April 1915. DOI: 10.5840/monist191525213.

[143] Hitti, P. K., *History of the Arabs*, Macmillian & Company Ltd., New York, 1963. DOI: 10.1007/978-1-137-03982-8.

[144] Hogendijk, Jan P. and Abdelhamid I. Sabra, *The Enterprise of Science in Islam*, The MIT Press, Cambridge, 2003.

[145] Hoggatt, V. E., *Fibonacci and Lucas Numbers*, Fibonacci Association, San Jose, 1973.

[146] Hooda, D. S. and J. N. Kapur, *Ārybhaṭa: Life and Contributions*, New Age International, New Delhi, 2nd ed., 2001.

[147] Horadam, A. F., Eight hundred years young, *The Australian Mathematics Teacher*, 31, 123–134, 1975.

[148] Horgan, John, Josephson's inner juction, *Scientific American*, 272(5), 40–42, May 1995. DOI: 10.1038/scientificamerican0595-40.

[149] Horne, R. A., Atomism in ancient Greece and India, *Ambix*, 8, 98–110, 1960. DOI: 10.1179/amb.1960.8.2.98.

[150] Horton, Rod W. and Herbert W. Edwards, *Backgrounds of American Literary Thought*, New York, 1952, first self published, later by Prentice Hall.

[151] Hughes, Barnabas, *Robert of Chester's Latin Translation of al-Khwarizmi's al-Jabr*, Franz Steiner Verlag, Stuttgart, 1989.

[152] Ifrah, Georges, *From One to Zero*, (Translator) Lowell Bair, Viking Penguin Inc., New York, 1985 (original in French).

[153] Iyengar, B. K. S., *Light on Yoga*, Schocken Books, New York, 1966.

[154] Jain, G. R., *Cosmology Old and New*, Bhartiya Jnanpith Publications, New Delhi, 1975.

[155] Josephson, Brian, Physics and spirituality: The next grand unification?, *Physics Education*, 22, 15–19, 1987. DOI: 10.1088/0031-9120/22/1/002.

[156] Joshi, Rasik Vihari, Notes on guru, Dīkṣā, and mantra, *Ethnos*, 37, 103–112, 1972. DOI: 10.1080/00141844.1972.9981054.

[157] Joshi, M. C. and S. K. Gupta, Eds., *King Chandra and the Meharauli Pillar*, Kusumanjali Prakashan, Meerut, 1989.

[158] Jung, C. G., *Letters*, G. Adler, Ed., Princeton University Press, 1973–75, 2 volumes.

[159] Jung, C. G., *The Collected Works of C. G. Jung*, Eds., H. Read, M. Fordham, and G. Alder, Princeton University Press, Princeton, 1954–1979, 20 volumes.

[160] Jungk, Robert, *Brighter Than a Thousand Suns*, HarCourt Brace, New York, 1958.

[161] Kak, S. C., Some early codes and ciphers, *Indian Journal of science*, 24, 1–7, 1989.

[162] Kak, S. C., The sign for zero, *The Mankind Quarterly*, 30, 199–204, 1990.

[163] Kapil, R. N., Biological sciences: Biology in ancient and medieval India, *Indian Journal History of Science*, 5(1), 119–140, 1970.

[164] Karpinski, Louis C., The Hindu-Arabic numerals, *Science*, 35(912), 969–970, June 21, 1912. DOI: 10.1126/science.35.912.969.

[165] Katz, Victor J., *History of Mathematics*, HarperCollins College Publishers, New York, 1993.

[166] Kaza, Stephanie, Western Buddhist motivations for vegetarianism, *Worldviews*, 9(3), 211–227, 2005. DOI: 10.1163/1568535057748416S0.

[167] Kennedy, E. S., The Arabic heritage in the exact sciences, *Quarterly Journal of Arab Studies*, 23, 327–344, 1970.

[168] Kennedy, E. S. and W. Ukashah, Al-Khwārizmī's planetary latitude tables, *Centaurus*, 14, 89–96, 1969. DOI: 10.1111/j.1600-0498.1969.tb00138.x.

[169] Khalidi, T., *Islamic Historiogaphy*, State University of New York Press, Albany, 1975.

[170] King, David, *Al-Khwārizmī and New Trends in Mathematical Astronomy in the Ninth Century*, Hagop Kevorkian Center for Near Eastern Studies, New York, 1983.

[171] King, D. A., *Astronomy in the Service of Islam*, Variorum, Aldenshot 1993. DOI: 10.1007/978-1-4614-6141-8_13.

[172] King, D. A. and G. Saliba, Eds., *From Deferent to Equant*, The New York Academy of Science, New York, 1987.

[173] King, Richard, *Indian Philosophy: An Introduction to Hindu and Buddhist Thought*, Georgetown University Press, Washington D.C., 1999.

[174] Krishnamurthy, K. H., *The Wealth of Suśruta*, International Institute of Ayurveda, Coimbatore, 1991.

[175] Krupp, E. C., Arrows in flight, *Sky and Telescope*, 92(4), 60–62, October 1996.

[176] Kulkarni, R. P., The value of π known to Śulbasūtrakarās, *Indian Journal of History of Science*, 13, 32–41, 1978a.

[177] Kulkarni, R. P., *Geometry According to Śulba Sūtra*, Vaidika Samsodhana Mandala, Pune, 1983.

[178] Kulkarni, T. R., *Upanishads and Yoga*, Bhartiya Vidya Bhavan, Bombay, 1972.

[179] Kumar, Alok, Improving science instruction by utilizing historical ethnic contributions, *Physics Education*, 11(2), 154–163, 1994.

[180] Kumar, Alok, *Sciences of the Ancient Hindus: Unlocking Nature in the Pursuit of Salvation*, CreateSpace, South Carolina, 2014.

[181] Kumar, Alok and Ronald A. Brown, Teaching science from a world-cultural point of view, *Science as Culture*, 8(3), 357–370, 1999. DOI: 10.1080/09505439909526551.

[182] Kunitzsch, P., How we got our Arabic star names, *Sky and Telescope*, 65, 20–22, 1983.

[183] Kunitzsch, P., *The Arabs and the Stars: Texts and Traditions on the Fixed Stars and their Influence in Medieval Europe*, Variorum, Northampton, 1989. DOI: 10.4324/9781315241340.

[184] Kutumbiah, P., *Ancient Indian Medicine*, Orient Longmans, New Delhi, 1962.

[185] Lad, Vasant, *Ayurveda: The Science of Self-healing, a Practical Guide*, Lotus Press, Santa Fe, 1984.

[186] Lahiri, Nayanjot, Indian metal and metal-related artifacts as cultural signifiers: An ethnographic perspective, *World Archaeology*, 27(1), 116–132, 1995. DOI: 10.1080/00438243.1995.9980296.

[187] Lattin, Harriet Pratt (Translator), *The Letters of Gerbert: With his Papal Privileges as Sylvester II*, Columbia University Press, New York, 1961.

[188] Le Coze, J., About the signification of Wootz and other names given to steel, *Indian Journal of History of Science*, 38(2), 117–127, 2003.

[189] Leeds, Anthony and Andrew Vayda, Eds., *The Role of Animals in Human Ecological Adjustments*, American Association for the Advancement of Science, Washington D.C., 1965.

[190] Levey, Martin and Marvin Petruck (Translator), *Principles of Hindu Reckoning*, by Kushyar Ibn Labban, University of Wisconsin Press, Milwaukee, 1965.

[191] Levey, Martin and Noury al-Khaledy, *The Medical Formulary of Al-Samarqandi*, University of Pennsylvania Press, Philadelphia, 1967. DOI: 10.9783/9781512803921.

[192] Little, A. G., Ed., *Roger Bacon*, Clarendon Press, Oxford, 1914.

[193] Loosen, Penate and Franz Vonnessen, Ed., *Gottfried Wilhelm Leibniz. Zwei Briefe über das Binäre Zahlensystem und die Chinesische Philosophie*, Belser-Presse, Stuttgart, 1968 (in German).

[194] MacKenna, Stephen, *Porphery on the Life of Plotinus* in Plotinus' Enneads, Faber and Faber Limited, London, 1956.

[195] Mahdihassan, S., Triphalā and its Arabic and Chinese synonyms, *Indian Journal of History of Science*, 13(1), 50–55, 1978.

[196] Majno, Guido, *The Healing Hand: Man and Wound in the Ancient World*, Harvard University Press, Cambridge, 1975. DOI: 10.1097/00006534-197602000-00022.

[197] Margenau, H., *Physics and Philosophy*, D. Reidel Publishing Company, London, 1978.

[198] Martzloff, Jean-Claude, *A History of Chinese Mathematics*, Springer-Verlag, New York, 1997. DOI: 10.1007/978-3-540-33783-6.

[199] Marx, Karl, The difference between the democritean and epicurean philosophy of nature, 1841, Doctoral Thesis, www.marxists.org/archive/marx/works/1841/dr-theses/

[200] McCall, Timothy B., *Yoga as Medicine: The Yogic Prescription for Health and Healing*, Bantam Books, New York, 2007.

[201] McCrindle, J. W., *Ancient India as Described by Megasthenes and Arrian*, Chuckervertty, Chatterjee and Company, Calcutta, 1926.

[202] McCrindle, J. W., *Ancient India as Described by Ktesias the Knidian*, Trubner and Company, London, 1882. Reprint by Manohar Reprints, Delhi, 1973.

[203] McDonell, John, J., *The Concept of an Atom from Democritus to John Dalton*, 11–12, The Edwin Mellen Press, Lewiston, 1991.

[204] McEvilley, Thomas, *The Shape of Ancient Thought*, Allworth Press, New York, 2002.

[205] Menninger, Karl, *Number Words and Number Symbols: A Cultural History of Numbers*, MIT Press, Massachusetts, 1970.

[206] Meyerhof, Max and Al at-Tabarî's "Paradise of Wisdom," one of the oldest Arabic compendiums of medicine, *ISIS*, 16(1), 6–54, 1931.

[207] Meyerhof, M., On the transmission of Greek and Indian science to the Arabs, *Islamic Culture*, 11, 17–29, 1937.

[208] Michalsen, Andreas, Gustav J. Dobos, and Manfred Roth, *Medicinal Leech Therapy*, Thieme Publishing, 2006. DOI: 10.1055/b-002-66250.

[209] Michalsen, A., U. Deuse, T. Esch, G. Dobos, and S. Moebus, Effect of leech therapy (*Hirudo Medicinalis*) in painful osteoarthritis of the knee: A pilot study, *Annals of the Rheumatic Diseases*, 60(6), 986, October 2001.

[210] Mikami, Yoshio, *The Development of Mathematics in China and Japan*, Chelsea Publishing Company, New York, 1913. DOI: 10.2307/3604893.

[211] Miller, Jeanine, *The Vision of Cosmic Order in the Vedas*, Routledge and Kegan Paul, Boston, 1985.

[212] Ming, Chen, The transmission of Indian ayurvedic doctrines in medieval China: A study of Aṣṭâṅga and Tridoṣa fragments from the silk road, *Annual Report of the International Research Institute for Advanced Buddhology at Soka University for the Academic Year 2005*, 9, 201–230, March 2006.

[213] Mishra, V. and S. L. Singh, Height and distance problems in ancient Indian mathematics, *Gaṇita Bhāratī*, 18, 25–30, 1996.

[214] Montgomery, Scott L. and Alok Kumar, *A History of Science in World Cultures: Voices of Knowledge*, Routledge, London, 2015. DOI: 10.4324/9781315694269.

[215] Montgomery, Scott L. and Alok Kumar, Telling stories: Some reflections on orality in science, *Science as Culture*, 9(3), 391–404, 2000. DOI: 10.1080/713695250.

[216] Moore, Arden and Sari Harrar, Leeches cut knee pain, *Prevention*, 55(8), 161, 2003.

[217] Moore, Walter, *Schrödinger: Life and Thought*, Cambridge Univesity Press, London, 1992.

[218] Mukherjee, Pulok K. and Atul Wahile, Integrated approaches towards drug development from Ayurveda and other Indian system of medicines, *Journal of Ethnopharmacology*, 103, 25–35, 2006. DOI: 10.1016/j.jep.2005.09.024.

[219] Mukherjee, P. K., *Indian Literature Abroad*, Calcutta Oriental Press, Calcutta, 1928.

[220] Mukhopādhyāya, Girinath, *Ancient Hindu Surgery*, Cosmo Publications, New Delhi, 1994, 2 volumes.

[221] Müller, M., *India: What can it Teach us?*, Funk and Wagnalls Publishers, London, 1883.

[222] Nakayama, S. and N. Sivin, Eds., *Chinese Science*, MIT Press, Massachusetts, 1973.

[223] Narayana, A., Medical science in ancient Indian culture with special reference to Atharvaveda, *Bulletin of the Indian Institute of History of Medicine*, 25(1–2), 100–110, 1995.

[224] Narayanan, Vasudha, Water, wood, and wisdom: Ecological perspectives from the Hindu traditions, *Daedalus*, 130(4), 179–197, 2001.

[225] Needham, J., *Science and Civilization in China*, Oxford University Press, London, 1954–1999, 6 volumes.

[226] Needham, Joseph, *Science in Traditional China: A Comparative Perspective*, Harvard University Press, Cambridge, 1981. DOI: 10.1119/1.13293.

[227] Needham, J. and W. Ling, *Science and Civilization in China*, Cambridge University Press, Chicago, 1959.

[228] Nelson, Lance, Ed., *Purifying the Earthly Body of God: Religion and Ecology in Hindu India*, State University of New York Press, Albany, 1998.

[229] Neugebauer, Otto E., *The Astronomical Tables of al-Khwārizmī*, Hist. Filos. Skr. Dan. Vid. Selsk, Copenhagen, 1962.

[230] Nicholson, Shirley J., Virginia Hanson, and Rosemarie Stewart, *Karma: Rhythmic Return to Harmony*, Motilal Banarsidass, New Delhi, 2002.

[231] Ninivaggi, Frank John, *Ayurveda: A Comprehensive Guide to Traditional Medicine for the West*, Praeger, Connecticut, 2008.

[232] Nisbet, Robert A., *Teachers and Scholars: A Memoir of Berkely in Depression and War*, Transaction Publishers, 1992. DOI: 10.4324/9781315130675.

[233] Oevi M., M. Rigbi, E. Hy-Am, Y. Matzner, and A. Eldor, A potent inhibitor of platelet activating factor from the saliva of the leech Hirudo medicinalis, *Prostaglandins*, 43, 483–95, 1992. DOI: 10.1016/0090-6980(92)90130-l.

[234] Ōhashi, Yukio, Astronomical instruments in classical Siddhāntas, *Indian Journal of History of Science*, 29(2), 155–313, 1994.

[235] Panikkar, R., Time and history in the tradition of India: Kala and Karma, in the book *Culture and Time*, The UNESCO Press, Paris, 1976.

[236] Paramhans, A. A., Astronomy in ancient India—its importance, insight and prevalence, *Indian Journal of History of Science*, 26(1), 63–70, 1991.

[237] Patton, Laurie L., *Authority, Anxiety, and Canon: Essays in Vedic Interpretation*, SUNY Press, Albany, 1994.

[238] Pellat, C., *The Life and Works of Al-Jahiz*, University of California, Los Angeles, 1969.

[239] Peters, Christian J., Jamie Picardy, Amelia F. Darrouzet-Nardi, Jennifer L. Wilkins, Timothy S. Griffin, Gary W. Fick, Carrying capacity of U.S. agricultural land: Ten diet scenarios, *Elementa: Science of the Anthropocene*, 4, 1–15, 2016. DOI: 10.12952/journal.elementa.000116.

[240] Philostratus, *Life of Apollonius*, C. P. Jones (Translator), Penguins Books, New York, 1970; *The Life of Apollonius*, Translated by F. C. Conybeare, Harvard University Press, Cambridge, 1960 (first published in 1912).

[241] Pillai, A. K. B., *Transcendental Self: A Comparative Study of Thoreau and the Psycho-Philosophy of Hinduism and Buddhism*, University Press of America, Lanham, 1985.

[242] Pingree, D., *The Thousands of Abu Masher*, Warburg Institute, London, 1968.

[243] Pothula, V. B., T. M. Jones, and T. H. J. Lesser, Ontology in ancient India, *The Journal of Laryngology and Otology*, 115, 179–183, March 2001. DOI: 10.1258/0022215011907091.

[244] Prakash, B. and K. Igaki, Ancient iron making in Bastar district, *Indian Journal History of Science*, 19, 172–185, 1984.

[245] Prakash, Om, *Food and Drinks in Ancient India*, Munshi Ram Manohar Lal, Delhi, 1961.

[246] Prasad, Hari Shanker, Ed., *Time in Indian Philosophy*, Sri Satguru Publications, Delhi, 1992.

[247] Radhakrishna, B. P. and L. C. Curtis, *Gold: The Indian Scene*, Geological Society of India, Bangalore, 1991.

[248] Radhakrishnan, Sarvepalli, *Indian Philosophy*, Macmillan, New York, 1958, 2 volumes. DOI: 10.1093/mind/xxxvii.145.130.

[249] Rados, Carol, Beyond bloodletting: FDA gives leeches a medical makeover, *FDA Consumer*, p. 9, September-October, 2004.

[250] Rajgopal, L., G. N. Hoskeri, G. S. Seth, P. S. Bhulyan, and K. Shyamkishore, History of anatomy in India, *Journal of Postgraduate Medicine*, 48(3), 243–5, 2002.

[251] Ramakrishnappa, K., *Impact of Cultivation and Gathering of Medicinal Plants on Biodiversity: Case Studies from India*, FAO, Rome, 2002. `http://www.fao.org/DOCREP/005/AA021E/AA021e00.htm`

[252] Rana, R. E. and B. S. Arora, History of plastic surgery in India, *Journal of Postgrad. Med.*, 48, 76–78, 2002.

[253] Rao, T. R. N. and Subash Kak, *Computing Science in Ancient India*, The Center for Advanced Computer Studies, University of Southwestern Louisiana, Lafayette, 1998.

[254] Rashed, Roshdi, Ed., *Encyclopedia of the History of Arabic Science*, Routledge, New York, 1996. DOI: 10.4324/9780203329030.

[255] Rawlinson, H. G., *Intercourse between India and the Western World*, Cambridge University Press, Cambridge, 1926. DOI: 10.2307/1842656.

[256] Ray, Praphulla Chandra, Chemical knowledge of the Hindus of old, *ISIS*, 2(2), 322–325, 1919.

[257] Ray, Priyada Ranjan, Chemistry in ancient India, *Journal of Chemical Education*, 25, 327–335, 1948. DOI: 10.1021/ed025p327.

[258] Ray, P. C., *A History of Hindu Chemistry*, Indian Chemical Society, 1956. It was first published in 1902. There are several editions of this book with different publishers.

[259] Remy, A. F. J., *The Influence of India and Persia on the Poetry of Germany*, Columbia University Press, New York, 1901.

[260] Renfro, Dave L., The Hindu method for completing the square, *The Mathematical Gazette*, 91(521), 198–201, July 2007. DOI: 10.1017/s0025557200181525.

[261] Restivo, Sal P., Parallels and paradoxes in modern physics and eastern mysticism: I-A critical reconnaissance, *Social Studies of Science*, 8(2), 143–181, 1978; Parallels and paradoxes in modern physics and eastern mysticism: II–A sociological perspective on parallelism, *Social Studies of Science*, 12(1), 37–71, 1982. DOI: 10.1177/030631282012001003.

[262] Riché, Pierre, *Gerbert d'Aurillac, le Pape de L'an mil*, Fayard, Paris, 1987 (in French).

[263] Riepe, Dale, *The Philosophy of India and its Impact on American Thought*, Thomas Press, Springfield, 1970.

[264] Roy, Mira, Environment and ecology in the *Rāmāyaṇa*, *Indian Journal of History of Science*, 40(1), 9–29, 2005.

[265] Royal, Denise, *The Story of J. Robert Oppenheimer*, St. Martin's, New York, 1969.

[266] Royle, J. F., *Antiquity of Hindo Medicine*, W. H. Allen and Company, London, 1837.

[267] Ruegg, D. Seyfort, Mathematical and linguistic models in Indian thought: The case of zero and Śûnatā, *Wiener Zeitschrift für die Kunde Südasiens und Archiv für Indische Philosophie*, 22, 171–181, 1978.

[268] Rusk, Ralph L., Ed., *The Letters of Ralph Waldo Emerson*, Columbia University Press, New York, 1939.

[269] Sachau, E. C., *Alberuni's India*, S. Chand & Company, Delhi, 1964. DOI: 10.4324/9781315012049.

[270] Sache, M., *Damascus Steel, Myth, History, Technology Applications*, Stahleisen, Düsseldorf, 1994.

[271] Saha, M. N. and N. C. Lahri, *History of the Calendar*, Council of Scientific and Industrial Research, New Delhi, 1992 (First published in 1955).

[272] Said, Edward W., *Orientalism*, Pantheon Books, New York, 1978.

[273] Said, Edward, *Culture and Imperialism*, Vintage Books, New York, 1993.

[274] Saidan, A. S., The development of Hindu-Arabic arithmetic, *Islamic Culture*, 39, 209–221, 1965.

[275] Saidan, A. S., *The Arithmetic of al-Uqlīdisī*, D. Reidel Publishing Company, Dordrecht, 1978.

[276] Salem, S. I. and A. Kumar, *Science in the Medieval World*, University of Texas Press, Austin, 1991.

[277] Saliba, George, *A History of Arabic Astronomy: Planetary Theories During the Golden Age of Islam*, New York University Press, New York, 1994.

[278] Samsó, J., *Islamic Astronomy and Medieval Spain*, Variorum, Aldenshot, 1994.

[279] Sankhyan, Anek R. and George H. J. Weber, Evidence of surgery in ancient India: Trepanation at Burzahom (Kashmir) over 4,000 years ago, *International Journal of Osteoachaelogy*, 11, 375–380, 2001. DOI: 10.1002/oa.579.

[280] Sarasvati Amma, T. A., *Geometry in Ancient and Medieval India*, Motilal Banarsidas, Delhi, 1979.

[281] Sarkar, Prasanta Kumar and Anand Kumar Chaudhary, Ayurvedic Bhasma: The most ancient application of nano medicine, *Journal of Scientific and Industrial Research*, 69, 901–905, 2010.

[282] Sarma, Nataraja, Diffusion of astronomy in the ancient world, *Endeavour*, 24(4), 157–154, 2000. DOI: 10.1016/s0160-9327(00)01327-2.

[283] Sarsvati, Svami Satya Prakash, *Founders of Sciences in Ancient India*, Govindram Hasaram, Delhi, 2 volumes, 1986.

[284] Sarton, G., *Introduction to the History of Science*, Carnegie Institute, Washington, 1927–1947, 3 volumes.

[285] Sarton, George, Decimal systems early and late, *Osiris*, 9, 581–601, 1950. DOI: 10.1086/368540.

[286] Schoff, Wilfred H., The eastern iron trade of the Roman empire, *Journal of the American Oriental Society*, 35, 224–239, 1915. DOI: 10.2307/592648.

[287] Schrödinger, Erwin, *My View of the World*, Cambridge University Press, England, 1964. DOI: 10.1017/cbo9781107049710.

[288] Scott, David, Rewalking Thoreau and Asia: "Light from the East" for "a Ver Yankee sort of oriental," *Philosophy East and West*, 57(1), 14–39, 2007. DOI: 10.1353/pew.2007.0011.

[289] Seal, Brajendranath, *The Positive Sciences of the Ancient Hindus*, Longmans, Green and Company, New York, 1915. This book has been reprinted several times, most recently in 1985 by Motilal Banarsidass, Delhi.

[290] Sedlar, J. W., *India and the Greek World*, Rowman and Littlefield, New Jersey, 1980.

[291] Sedlar, J. W., *India in the Mind of Germany*, University Press of America, Washington, D.C., 1982.

[292] Seidenberg, A., The ritual origin of geometry, *Archive for History of Exact Sciences*, 1, 488, 1962. DOI: 10.1007/bf00327767.

[293] Selin, H., Ed., *Encyclopedia of the History of Science, Technology, and Medicine in Non-Western Cultures*, Kluwer Academic Publishers, 1997. DOI: 10.1007/978-1-4020-4425-0.

[294] Selin, Helaine, *Astronomy Across Cultures: The History of Non-Western Astronomy*, Kluwer Academic Publishers, Boston, 2000.

[295] Sen, S. N., Influence of India science on other culture areas, *Indian Journal History of Science*, 5(2), 332–346, 1970.

[296] Sen, S. N. and A. K. Bag, *The Śulbasūtras of Baudhāyana, Āpastamba, Kātyana, and Mānava*, Indian National Science Academy, New Delhi, 1983.

[297] Sen, Tansen, Gautama Zhuan: An Indian astronomer at the Tang Court, *China Report*, 31(2), 197–208, 1995. DOI: 10.1177/000944559503100202.

[298] Shamasastry, R., *Kauṭilya's Arthśāstra*, Mysore Printing and Publishing House, Mysore, 1960, 4th ed., (first published in 1915).

[299] Sharma, Vijay Lakshmi and H. C. Bhardwaj, Weighing devices in ancient India, *Indian Journal of History of Science*, 24(4), 329–336, 1989.

[300] Sharma, Bhu Dev and Nabarun Ghose, *Revisiting Indus-Sarasvati Age and Ancient India*, World Association for Vedic Studies, 1998.

[301] Sherby, Oleg D. and Jeffrey Wadsworth, Damascus steel, *Scientific American*, 112–120, February 1985. DOI: 10.1038/scientificamerican0285-112.

[302] Shukla, Kripa Shankar and K. V. Sarma, *Āryabhaṭīya of Āryabhaṭa*, Indian National Science Academy, New Delhi, 1976.

[303] Siddiqi, M. Z., *Paradise of Wisdom* by Ali B. Rabban at-Ṭabarî, Berlin, 1928; reprinted by Hamdard Press, Karachi, 1981.

[304] Sigler, Laurence (Translator), *Liber Abaci by Leonardo Fibonacci*, Springer-Verlag, New York, 2002.

[305] Singh, L. M., K. K. Thakral, and P. J. Deshpande, Suśruta's contributions to the fundamentals of surgery, *Indian Journal History of Science*, 5(1), 36–50, 1970.

[306] Singh, Nand Lal, Ramprasad, P. K. Mishra, S. K. Shukla, Jitendra Kumar, and Ramvijay Singh, Alcoholic fermentation techniques in early Indian tradition, *Indian Journal of History of Science*, 45(2), 163–173, 2010.

[307] Singh, Permanand, The so-called Fibonacci numbers in ancient and medieval India, *Historia Mathematica*, 12, 229–244, 1985. DOI: 10.1016/0315-0860(85)90021-7.

[308] Sinha, Nandlal, *Vaisesika Sutras of Kanada*, Bhuvaneswari Ashram, Allahabad, 1911.

[309] Sinha, Braj M., *Time and Temporality in Sāṁkya-Yoga and Abhidharma Buddhism*, Munshiram Manoharlal Publishers, New Delhi, 1983.

[310] Sircar, D. C., *Inscriptions of Aśoka*, Ministry of Information and Broadcasting, Government of India, New Delhi, 1957.

[311] Smith, A. K. and C. Weiner, Eds., *Robert Oppenheimer*, Harvard University Press, 1980.

[312] Smith, Brian K., Classifying animals and humans in ancient India, *Man*, 26(3), 527–548, 1991. DOI: 10.2307/2803881.

[313] Smith, C. S., Damascus steel, *Science*, 216, 242–244, 1982.

[314] Smith, David Eugene and Louis Charles Karpinski, *The Hindu-Arabic Numerals*, Ginn and Company, Boston, 1911.

[315] Smith, David Eugene, *History of Mathematics*, Ginn and Company, New York, 1925, 2 volumes.

[316] Smith, Vincent A., *Aśoka, the Buddhist Emperor of India*, S. Chand & Company, Delhi, 1964.

[317] Somayaji, Dhulipala Arka , *A Critical Study of the Ancient Hindu Astronomy in the Light and Language of the Modern*, Karnatak University, Dharwar, 1971.

[318] Sorta-Bilajac, Iva and Amir Muzur, The nose between ethics and aesthetics: Sushruta's legacy, *Otolaryngology-Head and Neck Surgery*, 137, 707–710, 2007. DOI: 10.1016/j.otohns.2007.07.029.

[319] Srinivasan, Sharada, Metallurgy of zinc, high-tin bronze and gold in Indian antiquity: Methodological aspects, *Indian Journal of History of Science*, 51(1), 22–32, 2016.

[320] Srinivasiengar, C. N., *The History of Ancient Indian Mathematics*, World Press, Calcutta, 1967.

[321] Staal, Fritz, Greek and Vedic Geometry, *Journal of Indian Philosophy*, 27, 105–127, 1999.

[322] Steiner, Rudolf, *Goethe the Scientist*, O. D. Wannamaker (Translator), Anthroposophic Press, New York, 1950.

[323] Steinfeld, Henning, Pierre Gerber, Tom Wassenaar, Vincent Castel, Mauricio Rosales, and Cees de Haan, *Livestock's Long Shadow: Environmental Issues and Options*, Food and Agriculture Organization of the United Nations, Rome, Italy, 2006.

[324] Stodart J. and M. Faraday, On the alloys of steel, *Philosophical Transactions of the Royal Society of London, Ser. A*, 112, 253–70, 1822. DOI: 10.1098/rspl.1815.0182.

[325] Strabo, *The Geography of Strabo*, Horace Leonard Jones (Translator), Harvard University Press, Cambridge, 1959.

[326] Subak, S., Global environmental costs of beef production, *Ecological Economics*, 30, 79–91, 1999; DOI: 10.1016/s0921-8009(98)00100-1.

[327] Subbarayappa, B. V. and K. V. Sarma, *Indian Astronomy*, Nehru Centre, Bombay, 1985.

[328] Subbarayappa, B. V., *Science in India: A Historical Perspective*, Rupa Publications, New Delhi, 2013.

[329] Subbarayappa, B. V., An Estimate of the Vaiśesika Sūtra in the history of science, *Indian Journal of History of Science*, 2(1), 22–34, 1967.

[330] Suzuki, Jeff, *A History of Mathematics*, Prentice Hall, New Jersey, 2002.

[331] Swerdlow, N. M. and O. Neugebauer, *Mathematical Astronomy in Copernicus's De Revolutionibus*, Springer, New York, 1984. DOI: 10.1007/978-1-4613-8262-1.

[332] Takakusu, J. (Translator), *The Buddhist Religion by I-tsing*, Munshiram Manoharlal, Delhi, 1966 (first published 1896).

[333] Talbot, Michael, *Mysticism and New Physics*, Routledge, London, 1981.

[334] Teller, Edward, *Energy from Heaven and Earth*, W. H. Freeman and Company, San Francisco, 1979. DOI: 10.1119/1.12078.

[335] Teresi, Dick, *Lost Discoveries*, Simon and Schuster, New York, 2002.

[336] Thibaut, G., On the Śulva-Sūtra, *Journal Asiatic Society*, Bengal, India, 1875, p. 227.

[337] Thompson, C. J., *The Lure and Romance of Alchemy*, George G. Harrap and Company Ltd., London, 1932.

[338] Thomas, A. P., Yoga and cardiovascular function, *Journal of the International Association of Yoga Therapists*, 4, 39–41, 1993.

[339] Thoreau, H. D., *The Writings of Henry David Thoreau*, AMS Press, New York, 1906.

[340] Thoreau, Henry David, *Walden; or, Life in the Woods*, Dover, 1995. DOI: 10.5962/bhl.title.146169.

[341] Thurston, Hugh, Planetary revolutions in Indian astronomy, *Indian Journal of History of Science*, 35(4), 311–318, 2000.

[342] Thurston, Hugh, *Early Astronomy*, Springer Verlag, New York, 1994. DOI: 10.1007/978-1-4612-4322-9.

[343] Van Horn, Gavin, Hindu tradition and nature, *Worldviews*, 10(1), 5–39, 2006.

[344] Van Nooten, B., Binary numbers in Indian antiquity, *Journal of Indian Philosophy*, 21, 31–50, 1993. DOI: 10.1007/bf01092744.

[345] Veith, Ilza and Leo M. Zimmerman, *Great Ideas in the History of Surgery*, Norman Publishing, 1993.

[346] Voltaire, *The Works of Voltaire*, John Morley, William F. Fleming, and Oliver Herbrand Gordon Leigh, Eds., E. R. Du Mont, London, 1901.

[347] Wei-Xing, Niu, An inquiry into the astronomical meaning of Rāhu and Ketu, *Chinese Astronomy and Astrophysics*, 19(2), 259–266, 1995. DOI: 10.1016/0275-1062(95)00033-o.

[348] Weizman, Howard, More on India's sacred cattle, *Current Anthropology*, 15, 321–323, 1974.

[349] White, Andrew D., *A History of the Warfare of Science with Theology in Christendom*, D. Appleton and Company, New York, 1897 (reprinted by Prometheus, 1993). DOI: 10.1017/cbo9780511700804.

[350] Wright, W., *A Short History of Syriac Literature*, Philo Press, Amsterdam, 1966 (first pub. 1894).

[351] Yadav, B. S., Man Mohan, Eds., *Ancient Indian Leaps into Mathematics*, Birkhäauser, 2011. DOI: 10.1007/978-0-8176-4695-0.

[352] Yoke, Ho Peng, *Li, Qi and Shu: An Introduction to Science and Civilization in China*, Dover Publications, inc., New York, 1985.

[353] Zimmer, Henry, *Hindu Medicine*, The Johns Hopkins Press, Baltimore, 1948. DOI: 10.1097/00007611-195104000-00031.

[354] Worthington, Vivian, *The History of Yoga*, Routledge & Kegan Paul, 1982.

[355] Zukav, Gary, *The Dancing Wu-Li Masters*, Fontana/Collins, London, 1979.

[356] Zysk, Kenneth, *Asceticism and Healing in Ancient India: Medicine in the Buddhist Monastery*, Oxford University Press, New York 1991.

Author's Biography

ALOK KUMAR

Alok Kumar is a Distinguished Teaching Professor of physics at the State University of New York at Oswego. He was born and educated in India. Later, he taught at California State University at Long Beach and received the Meritorious Performance and Professional Promise Award for excellence in teaching and research in 1990. He has been teaching in the American higher education for about four decades. In Oswego, Kumar has received the Chancellor's Award for Excellence in Teaching, a life-time SUNY award, in 1997 and the President Award for Creative and Scholarly Activity or Research, a life-time award, in 2002. He is a fellow of the Alexander von Humboldt Foundation, Germany, and a NOVA/NASA fellow. Kumar is active in the fields of atomic physics, chemical physics, history of science, and science education.

He has about 75 peer-reviewed publications, and has authored/coauthored three books: (1) *Science in the Medieval World*, (2) *Sciences of the Ancient Hindus: Unlocking Nature in the Pursuit of Salvation*, and (3) *A History of Science in World Cultures: Voices of Knowledge*. All three books deal with the cultural heritage studies in science, including the non-Western cultures. Kumar believes that, to understand modern science, it is essential to recognize that many of the most fundamental scientific principles are drawn from knowledge amassed by ancient civilizations.

Kumar strongly believes that, in a gadget-filled world, scientific literacy is becoming an essential requirement for everyday life. It is the duty of a scientist to disseminate scientific knowledge to the general public. He has done so through articles and interviews in the popular media, making documentary films on archaeological sites that are rich in science, offering institutes for the underprivileged and underrepresented middle school students to pursue a career in science and technology, and lecturing about science for the general public. There are about 120 articles/reports about his activities in the popular media. This includes press releases from Reuters, the Press Trust of India, articles in *The Washington Post*, *Family Life*, *The Scientists*, *The Post Standard*, *The Palladium Times*, *India Abroad*, *India West*, *The South Asian Times*, *Hinduism Today*, *AramcoWorld*, *Organiser*, and radio interviews.

Index

Printed in the United States
by Baker & Taylor Publisher Services